Oxford Lecture Series in
Mathematics and its Applications 7

Series editors

John Ball   Dominic Welsh

OXFORD LECTURE SERIES
IN MATHEMATICS AND ITS APPLICATIONS

1. J. C. Baez (ed.): *Knots and quantum gravity*
2. I. Fonseca and W. Gangbo: *Degree theory in analysis and applications*
3. P. L. Lions: *Mathematical topics in fluid mechanics, Vol. 1: Incompressible models*
4. J. E. Beasley (ed.): *Advances in linear and integer programming*
5. L. W. Beineke and R. J. Wilson (eds): *Graph connections: Relationships between graph theory and other areas of mathematics*
6. I. Anderson: *Combinatorial designs and tournaments*
7. G. David and S. W. Semmes: *Fractured fractals and broken dreams*

Fractured Fractals and Broken Dreams

Self-Similar Geometry through Metric and Measure

Guy David
Université de Paris XI and Institut Universitaire de France

and

Stephen Semmes
Department of Mathematics, Rice University

CLARENDON PRESS • OXFORD
1997

Oxford University Press, Great Clarendon Street, Oxford OX2 6DP
Oxford New York
Athens Auckland Bangkok Bogota Bombay
Buenos Aires Calcutta Cape Town Dar es Salaam
Delhi Florence Hong Kong Istanbul Karachi
Kuala Lumpur Madras Madrid Melbourne
Mexico City Nairobi Paris Singapore
Taipei Tokyo Toronto Warsaw
and associated companies in
Berlin Ibadan

Oxford is a trade mark of Oxford University Press

Published in the United States by
Oxford University Press Inc., New York

A catalogue record for this book is available from the British Library

Library of Congress Cataloging in Publication Data
(Data applied for)

ISBN 0 19 850166 8

Typeset by the authors
Printed in Great Britain by
Bookcraft (Bath) Ltd
Midsomer Norton, Avon

PREFACE

Imagine that a perfectly self-similar fractal set is generated by a simple iteration and then subjected to bending, twisting, breaking, or corrosion. What remains of the initial structure?

Self-similarity is by now a familiar concept. One can repeat combinatorial recipes over and over again to obtain typically "fractal" limiting sets which look exactly the same at all scales and locations. Examples are commonly known and much studied, but often in a way which depends on the particular structure in question. Much less has been attempted by way of general definitions that would incorporate a broad range of phenomena into a single language.

We propose a concept of self-similarity here called *BPI* ("big pieces of itself"). The precise definition is given in the first chapter, and some characterizations and consequences are provided in Chapter 6. Roughly speaking, the definition asks that inside any pair of balls in the space there be pieces of *substantial proportion* which look almost alike in the sense of bilipschitz equivalence (i.e., up to bounded distortion of distances). This provides a framework in which to talk about "all" spaces with some self-similarity, and to make comparisons between them. It also provides a language in which to consider crucial features of a geometry without regard to accidents of a particular realization.

We do not mean to suggest that this framework is definitive – nothing ever will be – but it seems to be rather rich and flexible, and potentially opens up many interesting new opportunities for investigation.

In addition to the usual fractals one should keep more exotic spaces in mind. Examples of BPI spaces come from Heisenberg groups, asymptotic geometry of finitely generated groups, and constructions with doubling measures. (See Chapters 2, 4, and 16.) Doubling measures arise from Riesz products, Riemann surfaces, or more naive probabilistic considerations (as discussed in Chapter 16). A basic problem is to decide whether there is always something like a combinatorial pattern, a group, or a dynamical system behind any given BPI set.

Even in the case of perfectly self-similar sets it can be difficult to "see" precise combinatorial structure through primitive considerations of measure and metric, and this is a basic issue that we seek to address. It is still more problematic in the presence of singularities and distortions.

There are tools available for detecting approximately Euclidean behavior in sets despite singularities and distortions, through the notions of *rectifiability* and *uniform rectifiability*. (See Chapter 3 below.) The BPI condition can be seen as an extension of uniform rectifiability in which a simple model (like Euclidean geometry) is not provided in advance.

One would like to say when two different sets have essentially the same kind of structure. For this purpose there is a notion of *BPI equivalence*, defined in the first chapter and developed in Chapter 7. This condition is weaker than bilipschitz equivalence. (Think of having parts of your self-similar set break off.) BPI equivalence is an equivalence relation, and uniformly rectifiable sets are those BPI sets which are BPI equivalent to a Euclidean space. To ask about the existence of special rules or patterns behind the structure of a BPI space one can look for BPI-equivalent models which are more perfectly self-similar.

There are also natural relations between BPI spaces, of one space being more primitive than another. In Chapter 11 we discuss the concept of one BPI space "looking down" on another. BPI equivalence implies "look-down" equivalence (also discussed in Chapter 11), but the converse remains open. Some geometric consequences of one space looking-down on another are derived in Chapter 12, and a variety of special situations are discussed in Chapters 11 and 13. Related examples are provided in Chapter 14.

Instead of looking at individual BPI sets we can also think about *BPI geometries*, meaning equivalence classes of BPI sets. This leads to many natural questions, e.g., how many BPI geometries are there, how can they be deformed, to what extent are there plenty of continuous deformations or mostly a kind of discreteness, how does the ordering induced by looking down behave, etc. Special cases are treated in Chapters 15, 16, and 17.

We do not manage to go very far in this book towards definitive answers for any of these questions. There are many examples and concepts and basic facts, but no crisp theorems. The subject remains a wilderness, with no central zone, and many paths to try.

The lack of main roadways is also one of the attractions of the subject. It enjoys diverse connections with other aspects of geometry and analysis, including geometric measure theory, real analysis, bilipschitz and quasiconformal mappings, asymptotic behavior of groups and manifolds, and dynamical systems. We hope that the present text will be accessible to researchers in many areas, and also to people interested in entering the field. Many open problems emerge which are largely untouched, even for special cases and examples, and the general language leads to new perspectives and questions about classical constructions.

The prerequisites for this book are relatively modest, and consist mainly of basic knowledge of metric spaces and measure theory. While additional expertise would sometimes be helpful, basic definitions are recalled when needed, with adequate references to provide the reader with enough information to find out or work out what is needed at the moment. The reader may find the exposition [Se8] a helpful introduction to related aspects of geometry and analysis.

Much of this work was done during visits of the second author to the Institut des Hautes Etudes Scientifiques, to which the authors are very grateful. The second author was partially supported by the U.S. National Science Foundation.

CONTENTS

1

BASIC DEFINITIONS

In this chapter we define the notions of *BPI spaces* and *BPI equivalence*, which are central to our study of self-similar geometry.

For the record, a *metric space* is a nonempty set M together with a metric $d(x,y)$, which is to say a symmetric nonnegative function on $M \times M$ that vanishes exactly when $x = y$ and satisfies the triangle inequality

$$d(x,z) \leq d(x,y) + d(y,z) \tag{1.1}$$

for all $x, y, z \in M$. As usual, we write $B(x,r)$ for the open ball in M with center x and radius r, and $\overline{B}(x,r)$ for the closed ball. We shall write $B_M(x,r)$ when we need to make M explicit. We denote by $\operatorname{diam} E$ the diameter of a subset E of M, which is defined by $\operatorname{diam} E = \sup\{d(x,y) : x, y \in E\}$.

Next we define a weak notion of *homogeneity* for a metric space in terms of the distribution of its *mass*.

Definition 1.1 (Ahlfors regularity) *A metric space $(M, d(x,y))$ is said to be* (Ahlfors) regular of dimension d *(or simply* regular*) if it is complete, has positive diameter, and if there is a constant $C > 0$ so that*

$$C^{-1} r^d \leq H^d(B(x,r)) \leq C\, r^d \tag{1.2}$$

for all $x \in M$ and $0 < r \leq \operatorname{diam} M$, where H^d denotes d-dimensional Hausdorff measure on M.

Recall that d-dimensional Hausdorff measure H^d is defined as follows. Given $A \subseteq M$ and $\delta > 0$ we set

$$H^d_\delta(A) = \inf\{\textstyle\sum_j (\operatorname{diam} E_j)^d : \{E_j\} \text{ is a sequence of sets in } M \text{ which covers } A \text{ and satisfies } \operatorname{diam} E_j < \delta \text{ for all } j\} \tag{1.3}$$

and $H^d(A) = \lim_{\delta \to 0} H^d_\delta(A)$. This defines an outer measure on M which is additive when restricted to Borel sets. A good general reference for Hausdorff measures is provided by [Ma].

A set A is said to have Hausdorff dimension d if $H^s(A) = 0$ when $s > d$ and $H^s(A) = \infty$ when $s < d$. Ahlfors regularity determines the Hausdorff dimension d, and it does so in a uniform and scale-invariant manner.

In practice the following observation makes it easier to check and use Ahlfors regularity.

Lemma 1.2 *Let $(M, d(x,y))$ be a complete metric space, and suppose that μ is a Borel measure on M with the property that there are positive constants K and d such that*

$$K^{-1}\, r^d \leq \mu(B(x,r)) \leq K\, r^d \tag{1.4}$$

for all $x \in M$ and $0 < r \leq \operatorname{diam} M$. Then $(M, d(x,y))$ is regular with dimension d, and there is a constant C depending only on K and d so that $C^{-1}\mu(E) \leq H^d(E) \leq C\,\mu(E)$ for all Borel sets $E \subseteq M$.

We shall explain this in Section 5.3.

The next definition will be used to make comparisons between metric spaces.

Definition 1.3 (Lipschitz and bilipschitz mappings) *Let $(M, d(x,y))$ and $(N, \rho(u,v))$ be metric spaces. A mapping $f : M \to N$ is said to be* Lipschitz *if there is a constant C such that*

$$\rho(f(x), f(y)) \leq C\, d(x,y) \tag{1.5}$$

*for all $x, y \in M$. We might say that f is C-*Lipschitz *to make the constant explicit. We say that f is* bilipschitz *if there is a C so that*

$$C^{-1}\, d(x,y) \leq \rho(f(x), f(y)) \leq C\, d(x,y) \tag{1.6}$$

*for all $x, y \in M$, and again we might say that f is C-*bilipschitz *to be more explicit. Two spaces are said to be* bilipschitz equivalent *if there is a bilipschitz mapping of one onto the other.*

Thus Lipschitz mappings do not expand distances very much, while bilipschitz mappings do not expand or contract distances by more than a bounded factor. If $f : M \to N$ is K-Lipschitz, then

$$H^d(f(A)) \leq K^d\, H^d(A) \tag{1.7}$$

for all $A \subseteq M$. This follows directly from the definitions. Note that the two uses of H^d here are not quite the same, since one refers to M, the other to N. Similarly, if f is K-bilipschitz, then

$$K^{-d}\, H^d(A) \leq H^d(f(A)) \leq K^d\, H^d(A). \tag{1.8}$$

The next definition introduces terminology which is convenient but not standard.

Definition 1.4 (Conformally bilipschitz mappings) *Given a pair of metric spaces $(M, d(x,y))$ and $(N, \rho(u,v))$ and a mapping $f : M \to N$ between them, we say that f is C-*conformally bilipschitz *if there is a $\lambda > 0$ such that*

$$C^{-1}\lambda\, d(x,y) \leq \rho(f(x), f(y)) \leq C\,\lambda\, d(x,y) \tag{1.9}$$

for all $x, y \in M$. We call λ the scale factor. *(In practice we may say that f is C-*conformally bilipschitz with scale factor λ.*)*

Of course a conformally bilipschitz mapping is bilipschitz, but with a constant that depends on the scale factor λ. The reason for introducing this terminology is that we shall often try to compare pieces of metric spaces which lie within balls of very different radii, and we shall want uniform estimates which make sense independently of the choice of radii.

Note that the composition of two conformally bilipschitz mappings is conformally bilipschitz, with suitable behavior the constants.

The following is our basic notion of "self-similarity". It asks that for any pair of balls in the given space there be substantial subsets inside them which look approximately the same in terms of conformal bilipschitz equivalence.

Definition 1.5 (BPI spaces) *A metric space $(M, d(x, y))$ is a* BPI space *("big pieces of itself") if it is regular of some dimension d and if there exist constants $\theta, C > 0$ so that for each pair of balls $B(x_1, r_1)$ and $B(x_2, r_2)$ in M with $0 < r_1, r_2 \leq \operatorname{diam} M$ there is a closed set $A \subseteq B(x_1, r_1)$ with $H^d(A) \geq \theta\, r_1^d$ and a mapping $h : A \to B(x_2, r_2)$ which is C-conformally bilipschitz with scale factor r_2/r_1.*

Note that there is a uniform lower bound for $r_2^{-d}\, H^d(h(A))$ under the conditions above.

For us a ball $B(x, r)$ has always finite radius. It is sometimes convenient to write ranges of radii that theoretically permit $r = \infty$, as in the definition above, but whenever we write $B(x, r)$ we implicitly restrict ourselves to finite radii.

By asking only that pairs of balls contain substantial subsets which look alike, rather than whole replicas of each other, we have allowed a greater role for measure theory in this notion of self-similarity than is customary. We have increased the possibility for repetition, distortion, rupture, and clutter. The following notion of *BPI equivalence* takes these possibilities into account. It says that two BPI spaces are considered to be equivalent if for any ball in the first space and any ball in the second we can find substantial subsets which are conformally bilipschitz equivalent to each other.

Definition 1.6 (BPI equivalence) *Two BPI metric spaces $(M, d(x, y))$ and $(N, \rho(u, v))$ with the same dimension d are said to be* BPI equivalent *if there there exist constants $K, \alpha > 0$ so that for any $x \in M$, $0 < r \leq \operatorname{diam} M$, $u \in N$, and $0 < t \leq \operatorname{diam} N$ there exists a closed subset A of $B_M(x, r)$ with $H^d(A) \geq \alpha\, r^d$ and a K-conformally bilipschitz mapping $\phi : A \to B_N(u, t)$ with scale factor t/r.*

We shall see that BPI equivalence defines an equivalence relation on BPI spaces, and that two BPI spaces are BPI equivalent as soon as they contain a pair of subsets of positive measure which are bilipschitz equivalent. We can think of equivalence classes of BPI spaces as representing the same basic "geometry".

In Chapter 11 we shall discuss other relations between BPI spaces connected to the idea that one space is more "primitive" than another.

Our first task will be to give some examples BPI geometries that arise naturally in mathematics. Afterwards we establish basic facts about BPI spaces and

BPI equivalence. In Chapters 8 and 9 we develop the themes of compactness and convergence of metric spaces in connection with the BPI property and BPI equivalence.

2

EXAMPLES

2.1 Euclidean spaces

The first example is Euclidean space $\mathbf{R}^n$ equipped with the Euclidean metric. This is certainly Ahlfors regular of dimension n (Hausdorff measure coincides with Lebesgue measure in this case), and it is also BPI, because all balls are conformally isometric to each other.

A union of two n-planes in some $\mathbf{R}^m$ is also regular and BPI. In $\mathbf{R}^n$ let $\{x_i\}_i$ be a sequence of points such that $|x_i| = 2^{-i}$ for each i. Then

$$E = \{0\} \cup \bigcup_i \overline{B}(x_i, 2^{-i}) \tag{2.1}$$

is a regular set of dimension n which is BPI.

These two examples illustrate the kind of singularities, or failures of true self-similarity, that the BPI condition allows. On the other hand all of these sets are BPI equivalent to each other. This also illustrates the idea that one might have a BPI space that is not completely beautiful but for which there is a much nicer space that represents the same BPI equivalence class.

2.2 The snowflake functor

Let $(M, d(x,y))$ be a metric space. Given a real number $0 < s < 1$ consider the space $(M, d(x,y)^s)$. This is still a metric space, as is well known. We call this transformation of $(M, d(x,y))$ into $(M, d(x,y)^s)$ the *snowflake functor*.

It is important here that $s \leq 1$; if $s > 1$, then in general $D(x,y) = d(x,y)^s$ is only a *quasimetric*, which means that the triangle inequality should be weakened to

$$D(x,z) \leq K\,\{D(x,y) + D(y,z)\} \tag{2.2}$$

for some $K > 0$ and all $x, y, z \in M$. Conversely it turns out that if $D(x,y)$ is a quasimetric, then there exists a metric $d(x,y)$ and constants $C, s \geq 1$ such that

$$C^{-1}\,d(x,y)^s \leq D(x,y) \leq C\,d(x,y)^s \tag{2.3}$$

for all x and y. This is essentially the content of the proof of Theorem 2 in [MS].

The term *snowflake* here stems from the classical examples of snowflake curves in the plane, e.g., the Von Koch snowflake. When these constructions are made in a sufficiently regular manner they turn out to be bilipschitz equivalent to the standard line, interval, or circle (as appropriate), but with the metric deformed as in the snowflake functor for a suitable power s. The snowflake functor simply makes this idea abstract and applicable to any metric space. One should think of the classical snowflake pictures though, and their very crinkled shapes.

If $(M, d(x, y))$ is regular of dimension d, then $(M, d(x, y)^s)$ is regular of dimension d/s. If $(M, d(x, y))$ is BPI, then so is $(M, d(x, y)^s)$. This is an easy consequence of the definitions.

2.3 Cantor sets

Rather than discussing Cantor sets as explicitly constructed subsets of Euclidean spaces, we shall prefer to work with them in a more symbolic and abstract manner. Let F be a finite set with $k \geq 2$ elements. Let F^∞ denote the set of sequences $\{x_i\}_{i=1}^{\infty}$ with $x_i \in F$ for each i. F^∞ is our Cantor set, which we might call the k-*Cantor* set to be precise. It is a compact Hausdorff space when one gives it the usual product topology coming from the discrete topology on each factor of F.

We want our Cantor sets to be metric spaces, though. Given $x = \{x_i\}$ and $y = \{y_i\}$ in F^∞, let $L(x, y) = l$ if l is the largest integer such that $x_i = y_i$ when $1 \leq i \leq l$, and set $L(x, y) = \infty$ when $x = y$. ($L(x, y) = 0$ is allowed.) Given $0 < a < 1$ set

$$d_a(x, y) = a^{L(x,y)}, \tag{2.4}$$

where the right side is interpreted to be 0 when $L(x, y) = \infty$. This defines a metric on F^∞, and in fact it is an *ultrametric*, which means that

$$d_a(x, z) \leq \max\{d_a(x, y), d_a(y, z)\} \tag{2.5}$$

for all x, y, z. These metrics are all related by the snowflake functor, i.e., $d_a(x, y) = d_b(x, y)^s$ when $a = b^s$.

There is a natural probability measure μ on F^∞, defined as follows. Let ν denote the probability measure on F that assigns mass $1/k$ to each element of F. We take μ to be the infinite product of copies of ν.

Given $x = \{x_i\} \in F^\infty$ and a nonnegative integer j, the ball $B_a(x, a^j)$ with respect to the metric $d_a(x, y)$ consists of the points $y = \{y_i\} \in F^\infty$ such that $y_i = x_i$ when $i \leq j$. Thus

$$\mu(B_a(x, a^j)) = k^{-j}. \tag{2.6}$$

Using this it is easy to check that $(F^\infty, d_a(x, y))$ is regular of dimension d, where d is determined by the equation

$$a^d = k^{-1}. \tag{2.7}$$

Notice that we get all possible $0 < d < \infty$ by varying a in $(0, 1)$.

It is easy to see that each space $(F^\infty, d_a(x, y))$ is BPI. Again, like Euclidean spaces, these spaces are much better behaved than that, they have much more symmetry, but one can modify them and still get BPI spaces.

These symbolic Cantor sets are bilipschitz equivalent to the standard self-similar Cantor sets constructed in Euclidean spaces. The original middle-thirds set, for instance, is bilipschitz equivalent to $(F^\infty, d_a(x,y))$ with $k = 2$ and $a = 1/3$. This is easy to check.

We have described here the simplest and most classical geometries to put on Cantor sets, but they are not the only possibilities. Instead of defining a metric on F^∞ as above, using a single number a, we can use a function $a : F \to (0,1)$, as follows. Let $x = \{x_i\}$ and $y = \{y_i\}$ in F^∞ be given, and let $L(x,y)$ be as above. Define the distance between x and y to be

$$\prod_{i=1}^{L(x,y)} a(x_i). \tag{2.8}$$

This agrees with the previous definition when a is constant. One can check that this defines a metric on F^∞, in fact an ultrametric, in such a way that F^∞ is again Ahlfors regular and BPI.

There is a general recipe for "twisting" the geometry of Cantor sets that we shall discuss in detail later, and which will be shown to be exhaustive in a certain sense (see Chapter 16). It is not clear in general which of these geometries is BPI. Bilipschitz and BPI equivalence turn out to be tricky even in very self-similar situations.

2.4 Other fractals

This time we do construct our sets inside of the plane $\mathbf{R}^2$ for simplicity. Let Q denote the unit square $[0,1] \times [0,1]$ in $\mathbf{R}^2$. We can decompose Q into 9 (closed) subsquares Q_i in the obvious manner, so that the Q_i's have disjoint interiors and sidelength $1/3$.

Now suppose that we take Q and replace it with the union of all the Q_i's except the one in the center of Q. This produces a set K_1 which is the union of 8 squares. We can now apply this process to each of these squares, and then repeat the procedure indefinitely, to get a sequence of compact sets $\{K_j\}$ in the plane, where each K_j consists of 8^j squares of sidelength 3^{-j}, and where these squares have disjoint interiors. Set $K = \bigcap_j K_j$. This defines a compact subset of $\mathbf{R}^2$ which is called *the Sierpinski carpet* (see Figure 2.1).

There is a natural probability measure μ on K, which can be defined as follows. For each j let μ_j denote the probability measure on K_j which is just $(9/8)^j$ times Lebesgue measure. Then $\{\mu_j\}$ converges in the usual weak topology to a probability measure μ on K. One can also realize K as the continuous image of an 8-Cantor set F^∞ in the obvious manner, and obtain the measure from the one on the Cantor set. One of the main properties of our measure μ on K is that if S is one of the 8^j squares of which K_j is composed, then the piece of K inside S has mass 8^{-j}. It is not hard to check that K is regular of dimension d, where $3^d = 8$, and that K is BPI. (Each $S \cap K$ is an exact replica of the original.)

We can modify this construction to get a fractal tree instead. For this we replace the initial square Q with the union of the five squares among the Q_i's which consist of the four in the corners and the one in the middle. This also gives

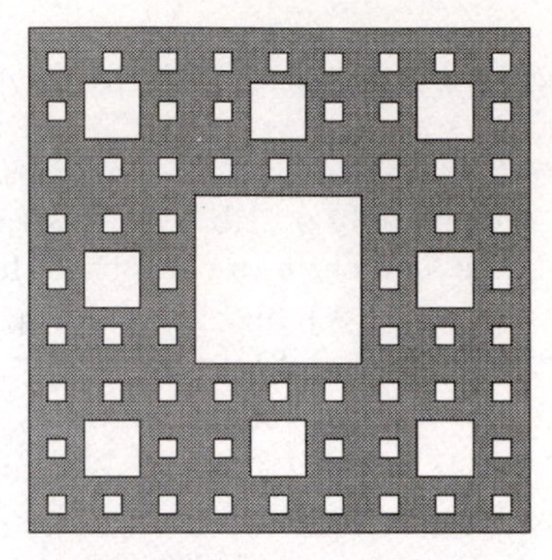

FIG. 2.1. The Sierpinski carpet

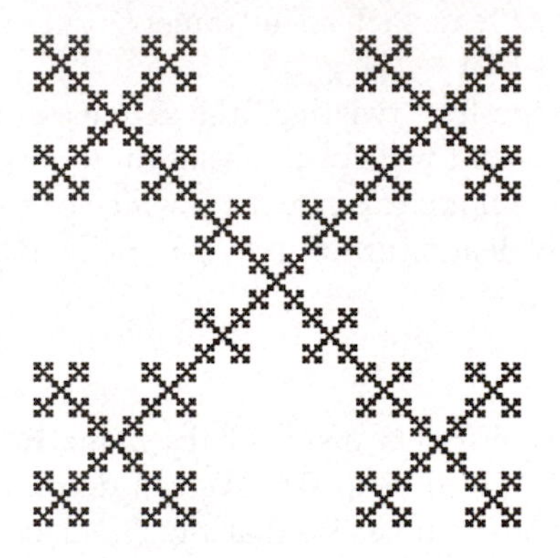

FIG. 2.2. A fractal tree

an Ahlfors regular BPI set of dimension d determined by $3^d = 5$. In this case the set is a fractal tree (see Figure 2.2).

Another example is given by the *Sierpinski gasket*. In this case we start with an equilateral triangle, we subdivide it into 4 subtriangles of equal size, and we keep the three in the corners and remove the one in the middle. This again leads to an Ahlfors regular BPI set (see Figure 2.3)

These examples are different in their geometry in terms of the number of ways that there are to connect points within the sets by curves of finite length. There are a lot more curves of finite length in the Sierpinski carpet than in the other two examples. There are many more rectifiable curves in Euclidean spaces of dimension > 1 than in the Sierpinski carpet, even if one accounts for the difference in the dimension. In Cantor sets there are no nontrivial curves whatsoever. In snowflakes there are plenty of nontrivial curves but no nontrivial rectifiable curves.

FIG. 2.3. The Sierpinski gasket

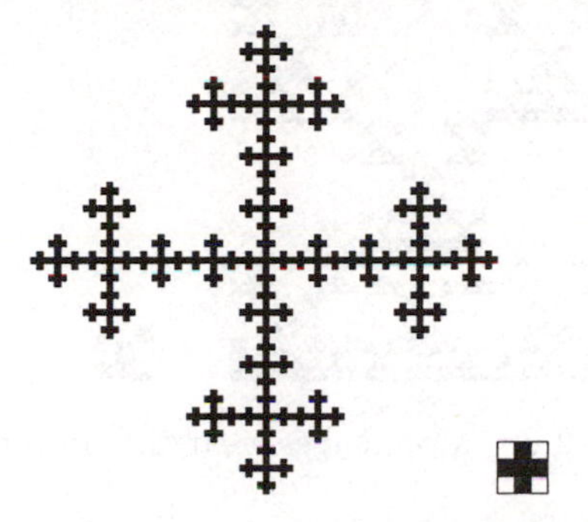

FIG. 2.4. Another fractal tree

2.5 A general procedure

Let Q denote the unit cube in $\mathbf{R}^n$. ("Cubes" should be interpreted as closed cubes here.) Fix an integer M, and subdivide Q into M^n subcubes of size M^{-1} in the obvious manner. Let $\mathcal{S}$ denote the collection of these subcubes.

Let $\mathcal{R}$ be a proper subset of $\mathcal{S}$. We think of $\mathcal{R}$ as being a "rule" for defining a (probably fractal) subset of Q. That is, we start with Q, we replace it with the cubes in $\mathcal{R}$, we identify each with Q in order to replace it with a collection of subcubes like $\mathcal{R}$, etc. If $\mathcal{R}$ has k elements, then this procedure produces a decreasing sequence of compact sets $\{A_j\}$, where A_j consists of k^j cubes of size M^{-j}. We take A to be the intersection of all the A_j's.

This procedure includes traditional Cantor sets and the examples of the previous section. Additional examples and their rules are given in Figures 2.4-2.8.

The limiting set A obtained in this way is Ahlfors regular of dimension d, where $M^d = k$. This is not hard to check. One can do the following, for instance. Let μ_j denote the probability measure on $\mathbf{R}^n$ supported on A_j which is uniformly distributed on A_j (with respect to Lebesgue measure). Thus μ_j assigns measure k^{-j} to each of the constituent cubes in A_j. It is not very difficult to see that $\{\mu_j\}$ converges weakly on $\mathbf{R}^n$ to a probability measure μ. If Q is one of the constituent cubes of A_j, then one can easily check that

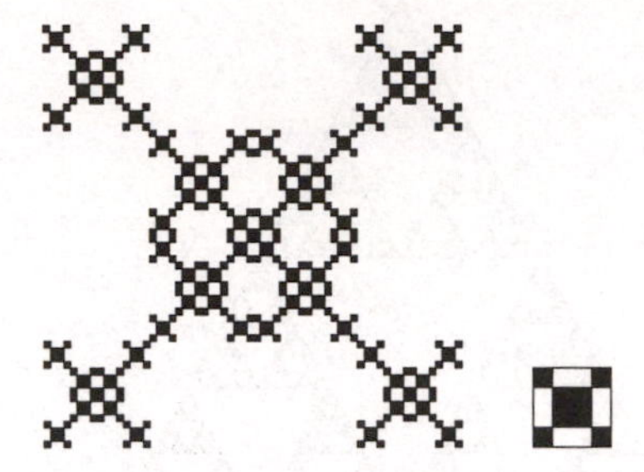

FIG. 2.5. More cubes in the middle

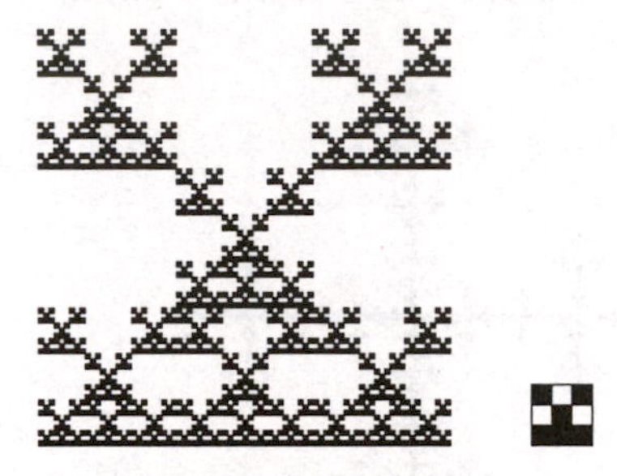

FIG. 2.6. More cubes near the bottom

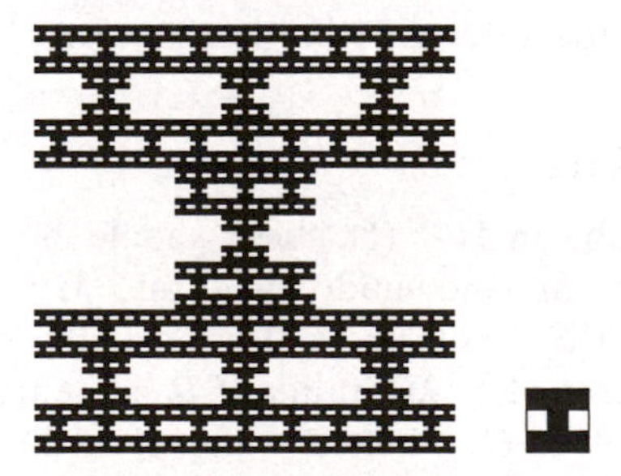

FIG. 2.7. Fractal train tracks

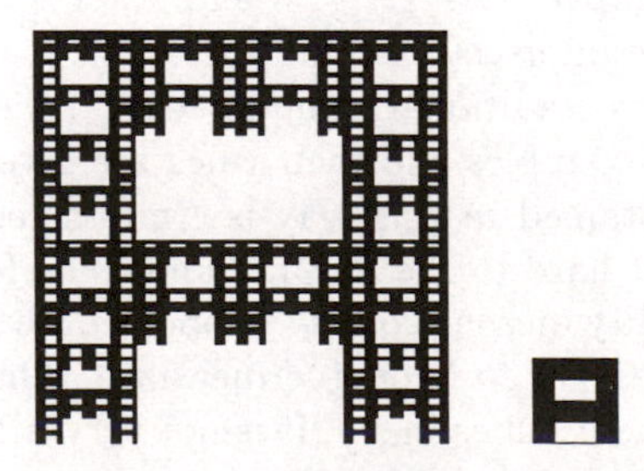

FIG. 2.8. The gates to knowledge

$$k^{-j} \leq \mu(Q) \leq C(n)\, k^{-j}, \tag{2.9}$$

where $C(n)$ depends only on n. In fact one can be careful and get $C(n) = 1$, but in related situations one has to think more about the possibility of mass accumulating on common faces of cubes.

This set is clearly BPI. Indeed, the set look the same modulo translations and dilations in each of the constituent cubes in A_j.

Fix M and k, and consider the collection of rules that we have described as defining a small universe of their own. This collection is typically not so small after all, since we are talking about choosing k elements from a set of cardinality M^n, and the number of ways of doing this can be quite large. Even with modest choices of the parameters there can be substantial variety in the phenomena that can occur, as indicated by the examples shown in this section and in the preceding one.

Let us think of this as a small universe of BPI sets and ask about the way that they compare with each other. Many rules will give practically the same result. In some cases this will be true in a very strong way, for instance if two rules are just translations of each other. In other cases it will be true in a more flexible manner. If a cube in a rule is isolated from the others, and it is moved to another position that is isolated from the others, then the change is not very important. There are many ways to get a set which amounts to a standard Cantor set. Even if cubes are not isolated from each other, it will frequently make no difference whether a chain of them points to the right or to the left.

One can formulate natural notions of combinatorial equivalence. A basic question then is to what extent the combinatorics can be recaptured from the metric geometry. One can make combinatorial criteria for two sets generated by these rules to be bilipschitz equivalent, but what about the converse? How much of the combinatorial structure of the rules has to be the same if the resulting sets are bilipschitz equivalent?

This is not so clear.

Here is a simple example. Take $n = 1$, $M = 5$, and $k = 3$. A rule consists of choosing three of the five equal pieces of the unit interval. Define one rule by taking the first, third, and fifth subintervals, and define a second rule by taking the first, fourth, and fifth. Consider the two sets that result from these two rules. Are they bilipschitz equivalent? We do not know. There is an obvious mapping from the first set onto the second which is Lipschitz but far from bilipschitz, but the absence of *any* bilipschitz mapping is not clear. We shall discuss this example more later, beginning in Section 11.6.

This example illustrates another point that we shall discuss later, the idea that one set can be more scattered than another, in the sense that there is a Lipschitz mapping from one onto the other. We shall discuss this further in Chapter 11. In our small universe of rules we see again that there can be simple combinatorial reasons for one of our sets to map onto another in a Lipschitz manner. For instance, if all the cubes in the rule are isolated from each other, then the resulting set can be mapped onto any other (in this universe) in a simple way. Again it is not clear what combinatorial conditions are necessary for the existence of such a mapping.

Thus in this small universe we can ask how the various rules are related combinatorially and in terms of the geometry of the sets that they generate. The collections of rules are rich enough to allow for some nontrivial relationships.

In this class of examples we have a nice illustration of the ideas of comparing BPI geometries, and having families of BPI geometries, rather than looking at BPI spaces as isolated objects.

One can also think about what happens when one builds sets by applying one rule after another, but maybe not the same rule each time. With a fixed choice of M and k, though, to have Ahlfors regularity anyway. This provides a way to build a large collection of sets in a way that one can describe – "parameterize" – and perhaps even analyze. It is natural to ask whether a set which is constructed in this manner and which has BPI should have some combinatorial structure behind it. We do not know the answer to this question. One can consider this question more generally for BPI spaces, but it is more crisp here.

We shall discuss these constructions further in Chapter 13.

2.6 Limit sets of discrete groups

For simplicity of exposition we shall permit ourselves to use more specialized terminology in this section without much explanation. We hope that the reader not acquainted with these topics will still get a flavor of the examples, for which we shall provide references.

Let $\mathbf{H}^{d+1}$ denote d-dimensional hyperbolic space, viewed as the unit ball in $\mathbf{R}^{d+1}$ equipped with the familiar metric. Let Γ be a discrete group of isometries on $\mathbf{H}^{d+1}$, which are Möbius transformations that can be extended also to the unit sphere $\mathbf{S}^d$. Under the "convex cocompact" assumption, Sullivan [Su1] has shown that the limit set of Γ is an Ahlfors regular set. (See also [Su2], especially Theorem 10.) One can use the convex cocompactness to show that the group visits all scales and locations on the limit set (this is better imagined in terms of the orbit of γ in hyperbolic space), in such a way that the geometry of Möbius transformations implies that the limit set is BPI. (In this case there is not a question of finding clever subsets of the space, one can just work with balls themselves.)

Sullivan's work was extended by Coornaert [Co] to groups acting on spaces of negative curvature in the sense of Gromov. In this case one has to be careful about what plays the role of the sphere at infinity, but there is a version of this. Under a convex-cocompactness assumption, Coornaert also establishes Ahlfors regularity of the limit set in the space at infinity. If we have understood this correctly, this limit set is BPI, and this method applies for instance to the case case of any compact negatively curved Riemannian manifold, acted on by the group of deck transformations. In this case the group is cocompact, and the limit set is the whole space at infinity.

3

COMPARISON WITH RECTIFIABILITY

3.1 Rectifiable sets in $\mathbf{R}^n$

Let d and n be integers, $0 < d < n$, fixed.

Definition 3.1 (Rectifiable sets) *A subset E of $\mathbf{R}^n$ is said to be rectifiable if there exists a sequence of subsets $\{E_j\}$ of E, each bilipschitz equivalent to a subset of $\mathbf{R}^d$, such that $H^d(E \setminus \bigcup_j E_j) = 0$.*

Strictly speaking we should say something like "rectifiable of dimension d", but it is simpler to let the dimension be understood implicitly.

Rectifiable sets are approximately Euclidean in their geometry. They need not be BPI, because no quantitative estimates are required, but the principle is similar. If E is measurable and rectifiable, then almost all points of E lie in an E_j, and therefore almost all points are points of density for some E_j. This means that there is a thick copy of a piece of a Euclidean space near the given point.

One of the nice features of rectifiability is that it is a very stable condition. Instead of asking that the E_j's be bilipschitz equivalent to subsets of $\mathbf{R}^d$ we could ask that they be Lipschitz images of subsets of $\mathbf{R}^d$, and we would get an equivalent condition. If instead we asked that they lie on C^1 manifolds of dimension d we would also get an equivalent condition. At bottom these equivalences come from the fact that Lipschitz functions on Euclidean spaces are differentiable almost everywhere. A convenient consequence of this fact is that a Lipschitz function f on a Euclidean space can be modified on a set of arbitrarily small measure to get a C^1 function.

These rigidity properties of Lipschitz mappings lead to stability properties of geometry, i.e., approximately Euclidean geometry in a weak sense implies approximately Euclidean behavior that appears to be much stronger. Covering almost everywhere by Lipschitz images of $\mathbf{R}^d$ implies covering almost everywhere by C^1 manifolds. Similarly, there are theorems which characterize rectifiability in terms of the requirement that for almost all points p in E the part of E near p lies mostly in a cone about a d-plane, and then this condition implies the apparently stronger property that the part of E near almost any point lies asymptotically close to a d-plane.

These stability properties help to make for nontrivial theorems about rectifiable sets. One can then ask whether similar phenomena hold for other kinds of geometries, such as the snowflakes and Cantor sets and fractals described in Chapter 2. The basic answer seems to be no, because of the absence of any rigidity theorems akin to differentiability almost everywhere. We say "seems" because it is not always clear how to formulate such rigidity theorems. Although the most obvious formulations fail, there does appear to be some room for structure. For instance, one might expect that there is more structure to a Sierpinski carpet because it has lots of rectifiable curves, even if there is not so much structure as for Euclidean spaces.

At any rate, in general one cannot expect results of the type "bounded equivalence implies asymptotic equivalence", as one has in the context of rectifiability and the differentiability of Lipschitz functions almost everywhere.

One reason for defining the concept of BPI sets is to provide a setting where we can ask ourselves such questions. Even if there are not so many rigidity results to be found, the lack of rigidity ought to have some interesting consequences for geometry. If we have a geometry which is not very rigid, then it should be easier to deform it, for example. One of the points of BPI is to try to have a language for formulating this idea in a precise way. More generally, one would like to be able to talk about different geometries at once, and moving between them.

Although this kind of incredibly strong rigidity does not work for most other spaces, there are other types of rigidity to look for, as we shall see. One can say that Euclidean geometry is incredibly un-rigid in terms of being able to map spaces into $\mathbf{R}^n$ easily, while it is very rigid in the opposite way, in the difficulty with mapping $\mathbf{R}^n$ into other spaces without having some very special structure. One can hope that there is some nontrivial "total" rigidity for other spaces, that the extent to which mappings on a given space do not have to have special structure is balanced by the special structure that is required for mappings into the space.

See [F1, Fe, Ma] for more information about rectifiability.

3.2 Uniform rectifiability

Uniform rectifiability is a variant of the notion of rectifiability which comes with uniform and scale-invariant estimates. Again let d and n be fixed integers with $0 < d < n$.

Definition 3.2 (Uniform rectifiability) *A subset E of $\mathbf{R}^n$ is said to be* uniformly rectifiable *if it is Ahlfors regular (of dimension d) and if there exist constants $M, \theta > 0$ so that for each $x \in E$ and $0 < r \leq \operatorname{diam} E$ there is a subset A of $E \cap B(x, r)$ such that*

$$H^d(A) \geq \theta\, r^d \tag{3.1}$$

and

$$A \text{ is } M\text{-bilipschitz equivalent to a subset of } \mathbf{R}^d. \tag{3.2}$$

This condition implies ordinary rectifiability (this is not to hard to show), but it is much stronger, because of the uniform bounds. This particular definition of uniform rectifiability should not be taken too seriously, in the sense that there are many equivalent conditions. See [D4, DS2, DS4, Se3] for more information. We could obviously extend the definition of uniform rectifiability to metric spaces instead of just subsets of Euclidean spaces, but some results about uniform rectifiability do not make sense in the abstract setting, and some are just not known. (Similar considerations apply to ordinary rectifiability.)

Proposition 3.3 *Uniformly rectifiable sets are BPI.*

To prove this we use the following.

Lemma 3.4 *Suppose that A_1, A_2 are two measurable subsets of the unit ball $B(0,1)$ in $\mathbf{R}^d$, with $|A_i| \geq \delta$ for some $\delta > 0$. (Here $|A|$ denotes the Lebesgue measure of A.) Then there is a point $z \in B(0,2)$ such that*

$$|\tau_z(A_1) \cap A_2| \geq C^{-1}\delta^2, \tag{3.3}$$

where $\tau_z : \mathbf{R}^d \to \mathbf{R}^d$ is the translation $\tau_z(x) = x - z$, and where C can be taken to be the volume of $B(0,2)$.

Indeed, Fubini's theorem implies that

$$\int_{B(0,2)} \left(\int_{B(0,1)} \mathbf{1}_{\tau_z(A_1)}(y)\, \mathbf{1}_{A_2}(y)\, dy \right) dz \geq \delta \int_{B(0,1)} \mathbf{1}_{A_2}(y)\, dy \geq \delta^2, \tag{3.4}$$

where $\mathbf{1}_A$ denotes the characteristic function of A. Hence

$$\int_{B(0,1)} \mathbf{1}_{\tau_z(A_1)}(y)\, \mathbf{1}_{A_2}(y)\, dy \geq \frac{\delta^2}{|B(0,2)|} \tag{3.5}$$

for some $z \in B(0,2)$, as desired.

The proposition is an easy consequence of the lemma. Indeed the lemma says that two bounded subsets of a Euclidean space of definite size always have subsets of definite size which are isometric (one is the image of the other by a translation). There is an obvious extension of this to subsets of balls of different size (with dilations used too), and the proposition follows from this extended version.

Of course we (the authors) began this story with uniform rectifiability and the question of whether there was something like it for other geometries. As in the comments at the end of the preceding section, we cannot expect anything like uniform rectifiability in its full glory for general BPI geometries, too much structure is missing. But again BPI provides a setting in which to ask certain questions, and to look at how geometries behave differently when one does not have so much structure as in uniform rectifiability.

4

THE HEISENBERG GROUP

The Heisenberg group is a nilpotent Lie group – or rather a family of Lie groups, one for each odd dimension ≥ 3 – which have many features in common with Euclidean spaces. Specifically, for each positive integer n we take H_n to be the set $\mathbf{C}^n \times \mathbf{R}$ equipped with the group operation

$$(z,t)\,(z',t') = \Bigl(z + z', t + t' + 2\,\mathrm{Im} \sum_{j=1}^{n} z_j\, \overline{z}_j'\Bigr). \tag{4.1}$$

This defines a noncommutative nilpotent group. There is a natural family of dilations on H_n given by $\delta_r : H_n \to H_n$ defined by

$$\delta_r(z_1, \ldots, z_n, t) = (r\, z_1, \ldots, r\, z_n, r^2\, t), \qquad r > 0, \tag{4.2}$$

and these dilations preserve the group structure.

This structure of group and dilations automatically determines a quasimetric on H_n. We start with the "homogeneous norm"

$$\|(z_1, \ldots, z_n, t)\| = \Bigl(\sum_{j=1}^{n} |z_j|^4 + |t|^2\Bigr)^{\frac{1}{4}}. \tag{4.3}$$

This has the invariance properties that

$$\|\delta_r(x)\| = r\, \|x\| \tag{4.4}$$

and

$$\|x^{-1}\| = \|x\| \tag{4.5}$$

for all $x \in H_n$ and $r > 0$, where x^{-1} denotes the inverse of x in H_n. (If $x = (z_1, \ldots, z_n, t)$, then $x^{-1} = (-z_1, \ldots, -z_n, -t)$.) For our purposes the specific form of $\|\cdot\|$ does not matter, what matters is that it satisfies these invariance properties, and that it is continuous and positive away from the origin. Any function on H_n with these properties will be bounded above and below by constant multiples of $\|\cdot\|$. We use this homogeneous norm to define a quasimetric on H_n by

$$d(x,y) = \|x^{-1}\, y\|. \tag{4.6}$$

This is symmetric because of (4.5), it is clearly nonnegative and vanishes exactly when $x = y$, and it is not hard to check that it satisfies the approximate triangle inequality (2.2).

This quasimetric has important symmetry properties. If $a \in H_n$ let $\lambda_a : H_n \to H_n$ denote the mapping of left translation by a (with respect to the group structure), so that $\lambda_a(x) = a\,x$. Our quasimetric $d(x, y)$ is invariant under left translations, $d(\lambda_a(x), \lambda_a(y)) = d(x, y)$ for all $x, y, a \in H_n$. It also behaves well under dilations,

$$d(\delta_r(x), \delta_r(y)) = r\,d(x, y) \tag{4.7}$$

for all $x, y \in H_n$ and all $r > 0$.

Notice that $d(x, y)$ is compatible with the Euclidean topology on H_n. Any other quasimetric on H_n which is compatible with the Euclidean topology and which has the same behavior under translations and dilations is practically the same as $d(x, y)$, i.e., is bounded above and below by constant multiples of $d(x, y)$.

Actually this quasimetric turns out to be a metric. See [KR] for the computation. The preceding discussion is more robust; even if one accidentally chooses a quasimetric which is not a metric but otherwise has all the right properties, then it is still bounded above and below by constant multiples of a metric, and is therefore practically as good as a metric.

To really understand the metric geometry of the Heisenberg group one should see it also as a subriemmanian geometry, with plenty of rectifiable curves and a natural notion of geodesics. We shall not go into this here.

To put the Heisenberg geometry into perspective one should consider its simpler cousin where we use the same homogeneous norm as in (4.3), but we set

$$D(x, y) = \|x - y\| \tag{4.8}$$

for all $x, y \in \mathbf{C}^n \times \mathbf{R}$. Here $x - y$ uses the ordinary Euclidean subtraction. This geometry still satisfies the scaling law (4.7), but the difference is that the "axes" are not turning. By turning the axes the Heisenberg geometry becomes more rich, e.g., every pair of points can be connected by a curve of finite length, which is not true for (4.8). The latter amounts to the Cartesian product of a $2n$-dimensional Euclidean space with a snowflake metric on the real line.

Both of these spaces are Ahlfors regular of dimension $2n+2$, as one can easily check using Lemma 1.2 and ordinary Lebesgue measure. They are both BPI, and in fact any pair of balls is conformally isometric, using translations and dilations.

The Heisenberg group has the amazing property that Lipschitz functions on it are differentiable almost everywhere. This means real-valued Lipschitz functions, and it also works for Lipschitz mappings in other Heisenberg groups (or "Carnot groups"). This is proved in [P1]. This notion of differentiability is defined by using the translations and dilations to blow up a mapping at a point, in the same manner as for real-valued functions on Euclidean spaces. Note that we need translations and dilations on both the target space and the domain to make this blowing up. Pansu proves then that when you blow up a Lipschitz mapping at almost any point, then the limit exists, and the limiting mapping is a group homomorphism. The group homomorphisms play the role that affine mappings have for Euclidean spaces. (Of course one should normalize the blow-ups to be mappings which take the origin to the origin.)

In view of this differentiability almost everywhere one should expect a good theory of rectifiability modelled on the Heisenberg group just as there is for ordinary rectifiability and Euclidean geometry, as in Chapter 3. No one seems to have worked this out, let alone the corresponding version of uniform rectifiability. Still, in this case, unlike other fractals, it is reasonable to expect a theory much like the Euclidean one.

With some limitations, though. Part of the strength of the Euclidean theories of rectifiability and uniform rectifiability is their connection with topology. The equality of Hausdorff dimension and topological dimension often leads to some nontrivial geometric structure like rectifiability. The Heisenberg group, by contrast, has Hausdorff dimension 1 larger than the topological dimension. For Euclidean geometry there are criteria for rectifiability and uniform rectifiability which are based on quantitative topological conditions, as in [D1, D3, D4, DJ, DS3, DS5, DS6, Se3]. Part of the stability of "approximate Euclidean geometry" seems to come from this relation with topology, as though topology provides a floor through which one cannot pass and which then leads to a lot of structure in the presence of upper bounds on the mass (as in the "Second Principle" articulated in [Se3]). There is nothing apparently like this for the Heisenberg group, and so one might expect some of the same sort of stability of geometry as one has for Euclidean geometry but still not as much. (See also [Se7].)

5

BACKGROUND INFORMATION

5.1 Extending Lipschitz functions

Let $(M, d(x,y))$ be a metric space, and let E be a nonempty subset of M. Suppose that we have a Lipschitz mapping $f : E \to \mathbf{R}$. Then we can extend f to a Lipschitz function on all of M, and with the same Lipschitz constant. This is well known, but let us quickly recall the proof. Suppose that f is L-Lipschitz. Define $g : M \to \mathbf{R}$ by

$$g(x) = \inf\{f(y) + L\, d(x,y) : y \in E\}. \tag{5.1}$$

For each $y \in E$, the function $x \mapsto f(y) + L\, d(x,y)$ is an L-Lipschitz function, as one can easily check. (The mapping $x \mapsto d(x,y)$ is 1-Lipschitz, because of the triangle inequality.) We have that $g \geq f$ on E automatically, and the Lipschitz condition for f implies that $g = f$ on E. It is easy to check that g is finite everywhere and L-Lipschitz.

5.2 A covering lemma

Let $(M, d(x,y))$ be a metric space, let E be a subset of M, and let $\mathcal{B}$ be a collection of balls in M of bounded radii which covers E. Then there is a sequence $\{B_j\}$ of balls in $\mathcal{B}$ such that the elements of $\{B_j\}$ are pairwise disjoint, and either there are infinitely many B_j's and

$$\limsup_{j \to \infty} \operatorname{radius} B_j > 0, \tag{5.2}$$

or we have $E \subseteq \bigcup_j 5\, B_j$. This is a well-known covering lemma of Vitali type, as in [St1]. It is proved by a simple greedy method; at each stage one chooses balls disjoint from the ones already chosen and which are as large as they can be, to within a factor of 2.

If E is bounded and $(M, d(x,y))$ is Ahlfors regular, then one has that the radii of the B_j's tend to 0 automatically, because the sum of their measures must be finite, and hence their measures tend to 0.

This kind of covering argument works no matter whether one uses closed balls or open balls.

5.3 Ahlfors regular spaces

Lemma 5.1 *Let $(M, d(x,y))$ be a regular metric space of dimension d (Definition 1.1). Then for each ball B in M of radius $R \leq \operatorname{diam} M$ and each $0 < r < R$*

we can cover B by $\leq C\,(R/r)^d$ balls of radius r, where C depends only on the regularity constant for M.

Indeed, let B, R, r be given as above, and let A be a subset of B all of whose elements are at a distance $\geq r$ from each other. Then the balls $B(a, r/2)$ are pairwise disjoint and contained in $2\,B$. Using Ahlfors regularity we can convert the fact that the sum of the measures of the balls $B(a, r/2)$, $a \in A$, is no greater than the measure of $2\,B$ into the fact that A has at most $C\,(R/r)^d$ elements.

Since we have a bound on the possible number of elements of such a set A, we can take one with a maximal number of points. In this case B will be covered by the balls $B(a, r/2)$, $a \in A$, by maximality, and the lemma follows.

Corollary 5.2 *Closed and bounded subsets of an Ahlfors regular space are compact.*

Indeed, bounded subsets are "totally bounded" (they can be covered by finitely many balls of arbitrarily small radius, by the preceding lemma), and since regular spaces are assumed to be complete we get compactness by a well-known fact from real analysis.

The preceding lemma and corollary apply equally well if we assume that $(M, d(x, y))$ satisfies the assumptions of Lemma 1.2 instead of Ahlfors regularity. One conclusion of this observation is that if μ is as in Lemma 1.2, then μ is Borel regular, i.e., the μ-measure of a Borel set is the infimum of the μ-measures of the open sets which contain the given set. This follows from Theorem 2.18 in [R], which requires that M be locally compact and that every open subset of M be a countable union of compact sets, which are both true in our case because of the corollary.

If μ is as in Lemma 1.2, then it is very easy to see that μ is bounded by a constant times H^d on Borel sets, just by tracking down the definitions. Conversely, the H^d-measure of a ball of radius R will be bounded by a constant times R^d because of the lemma above. One can extend this to show that the μ-measure of an open set is bounded by a constant multiple of the H^d-measure using standard covering arguments. The aforementioned Borel regularity then implies a similar inequality for all Borel sets. This is why Lemma 1.2 is true.

A lot of the standard real analysis on Euclidean spaces works on regular spaces, as discussed in [CW1, CW2]. For our purposes a key fact is that the usual story of Lebesgue points and points of density hold. That is, if $(M, d(x, y))$ is an Ahlfors regular space of dimension d and f is a locally integrable function on M, then

$$\lim_{r \to 0} \frac{1}{|B(x, r)|} \int_{B(x,r)} |f(y) - f(x)|\, dy = 0 \tag{5.3}$$

for almost all x. Here we begin to use the shorthand notation $|E|$ for the measure of a subset E of an Ahlfors regular space M (with respect to H^d or more generally any measure μ as in Lemma 1.2, it does not really matter), and we let dy, etc., denote the same measure for the purposes of writing integrals. (We shall frequently use these conventions.) Similarly, if E is a Borel subset of M, then

$$\lim_{r\to 0} \frac{|E \cap B(x,r)|}{|B(x,r)|} = 1 \tag{5.4}$$

at almost all points in E and the limit exists and equals 0 for almost all points in the complement of E. These facts are proved in essentially the same manner as for Euclidean spaces, as in the first few pages of [St1]. The main point is that the covering argument still works, as in the preceding section. See [CW1, CW2] for more information, and see also [St2].

The next definition makes precise the idea that a metric space does not have islands which are too isolated.

Definition 5.3 (Uniformly perfect spaces) *Let $(M, d(\cdot,\cdot))$ be a metric space. We say that M is* uniformly perfect *if there is a constant $C > 0$ such that for each $x \in M$ and $0 < t \leq \operatorname{diam} M$ there is a point $y \in M$ which satisfies $C^{-1} t \leq d(x,y) \leq t$.*

Lemma 5.4 *Ahlfors regular spaces are uniformly perfect.*

This is an easy exercise. ($B(x,r)\backslash B(x, C^{-1} r)$ has positive measure when C is large enough.)

5.4 Assouad's embedding theorem

Definition 5.5 (The doubling property) *A metric space M is said to be* doubling *if there is a constant C such that every ball B in M can be covered with at most C balls of half the radius of B.*

Ahlfors regular spaces are always doubling.

It is not hard to check that a subspace of a doubling metric space is also doubling. Thus any subspace of a Euclidean space is doubling. Is every metric space which is doubling bilipschitz equivalent to a subset of a Euclidean space? It turns out that if $(M, d(x,y))$ is doubling and $0 < s < 1$ then $(M, d(x,y)^s)$ is bilipschitz equivalent to a subset of some $\mathbf{R}^n$, where n and the bilipschitz constant depend only on s and the doubling constant. This was proved by Assouad [A1, A2, A3]. Notice that the doubling condition is necessary for such an embedding to exist. The Heisenberg group with the metric described in Chapter 4 provides an example of a metric space which is doubling but for which we cannot take $s = 1$. This was known to Assouad, and it follows from the differentiability almost everywhere of Lipschitz functions on the Heisenberg group, as discussed in [Se4].

Remember that any metric space $(M, d(x,y))$ can be isometrically embedded into a Banach space. In fact, if M is separable – which doubling spaces are – then M can be embedded isometrically in ℓ^∞. We take a dense sequence $\{x_j\}$ in M, we fix a basepoint p in M, and we define the mapping from M into ℓ^∞ by

$$x \in M \mapsto \{\alpha_j\}_j = \{d(x,x_j) - d(p,x_j)\}_j \in \ell^\infty. \tag{5.5}$$

That this is an isometric embedding is a well known and easy exercise.

The existence of these isometric embeddings means that if we want to talk about the set of all separable metric spaces, we can simply look inside the subsets of ℓ^∞.

5.5 Dyadic cubes

The usual decompositions of Euclidean spaces into dyadic cubes are very useful sometimes, and we would like to have a version for general Ahlfors regular spaces. Let $(M, d(x,y))$ be an Ahlfors regular metric space of dimension d, and assume for the moment that M is unbounded. Then there exists a family $\Delta_j, j \in \mathbf{Z}$, of Borel measurable subsets of M, with the following properties:

$$\begin{gathered}\text{each } \Delta_j \text{ is a partition of } M\text{, i.e., } M = \bigcup_{Q \in \Delta_j} Q \\ \text{and } Q \cap Q' = \varnothing \text{ whenever } Q, Q' \in \Delta_j \text{ and } Q \neq Q'; \end{gathered} \tag{5.6}$$

$$\begin{gathered}\text{if } Q \in \Delta_j \text{ and } Q' \in \Delta_k \text{ for some } k \geq j, \\ \text{then either } Q \subseteq Q' \text{ or } Q \cap Q' = \varnothing; \end{gathered} \tag{5.7}$$

$$\begin{gathered}\text{for all } j \in \mathbf{Z} \text{ and all } Q \in \Delta_j \text{ we have that} \\ C^{-1}\, 2^j \leq \operatorname{diam} Q \leq C\, 2^j \text{ and } C^{-1}\, 2^{jd} \leq |Q| \leq C\, 2^{jd}; \end{gathered} \tag{5.8}$$

$$\begin{aligned}&\text{for all } j \in \mathbf{Z} \text{ and } Q \in \Delta_j,\ Q \text{ has small boundary,} \\ &\text{in the sense that} \\ &\quad |\{x \in Q : \operatorname{dist}(x, E \backslash Q) \leq \tau\, 2^j\}| \leq C\, \tau^{1/C}\, 2^{jd} \\ &\text{and} \\ &\quad |\{x \in M \backslash Q : \operatorname{dist}(x, Q) \leq \tau\, 2^j\}| \leq C\, \tau^{1/C}\, 2^{jd} \\ &\text{for all } 0 < \tau < 1. \end{aligned} \tag{5.9}$$

The constant C here depends on the dimension d and on the regularity constant for M. Note that these properties hold for the standard decomposition of Euclidean spaces into dyadic cubes.

A proof of the existence of such a decomposition of a regular space can be found in [D4] when the space is contained inside a Euclidean space. The general case follows from Assouad's embedding theorem. See also [Ch] for a direct construction on general spaces.

In the case where M is bounded we have a similar decomposition, except that we should restrict ourselves to $j \in \mathbf{Z}$ which satisfy $j \leq j_0$, where $2^{j_0} \leq \operatorname{diam} M \leq 2^{j_0+1}$. We can derive the existence of this decomposition for bounded spaces from the corresponding result for unbounded spaces. We simply build a new metric space N from M in such a way that N consists of a sequence of copies of M, each with the metric dilated by a factor of $2^l, l = 0, 1, 2, \ldots$, and with the distances between any two of these copies being equal to the maximum of their diameters. It is easy to build such a space N.

We shall use these decompositions of regular spaces into cubes frequently. Let us mention a couple of simple facts about them. Let M be a regular metric space with a decomposition into cubes as above, and let Q denote one of these cubes. Then there is a point $x_Q \in Q$ such that

$$B(x_Q, C^{-1} \operatorname{diam} Q) \subseteq Q. \tag{5.10}$$

Here C is a constant that does not depend on Q. The existence of this kind of "central point" follows easily from (5.9). Indeed, for C large enough the points in Q for which this inclusion fails all lie close to the boundary, and so this set of points has small measure by (5.9). If we choose the constant large enough then there are still some points remaining in Q because of (5.8).

Another fact which is quite useful is that if Q is a cube in M then $\overline{Q}$ is Ahlfors regular of dimension d, with a constant that is bounded uniformly in Q. Completeness and the upper bound for the measures of balls in Q follow immediately from the corresponding statements for M, but for the lower bound we should be more careful. Let $x \in \overline{Q}$ and $0 < r \leq \operatorname{diam} Q$ be given, and let us check that

$$|B(x,r) \cap Q| \geq C^{-1} r^d. \tag{5.11}$$

Pick a point $y \in Q$ such that $d(x,y) < r/2$, and let Q' be the largest cube contained in Q which contains y and satisfies $\operatorname{diam} Q' \leq r/2$. It is not hard to show that $\operatorname{diam} Q' \geq C^{-1} r$, using the properties of our families of cubes. We also have $Q' \subseteq B(x,r)$, and so the desired estimate follows from (5.8) applied to Q'.

This fact makes it easy to "localize" Ahlfors regular spaces in a convenient way, which is not so obvious otherwise.

A very simple consequence of (5.9) is that the boundary of Q (with respect to the topology of M) always has measure 0. There are sometimes annoying ambiguities coming from the boundaries of cubes, but they do not really matter, because of this fact. For instance, we claim that if Q is a cube in M, then we can get a good family of cubes in $\overline{Q}$, viewed as a regular space in its own right, by adding sets of measure 0 to the boundaries of the subcubes of Q that come from M. This is not very difficult to achieve, adding points generation by generation, being careful to respect the compatibility required in (5.7).

The notion of cubes will be very convenient for studying BPI spaces. Let us make a couple of observations about that now. If we take the definition of BPI spaces and replace the balls with cubes, then we get an equivalent condition. That is, if we require that for any pair of cubes we can find subsets of them of substantial proportion which are conformally bilipschitz with uniform bounds, then we get an equivalent condition. This follows from the basic properties of cubes, and in particular the aforementioned existence of a central point.

If now M is a BPI space and Q is a cube in M, then $\overline{Q}$ is also BPI, with a constant that does not depend on Q. We saw above that $\overline{Q}$ is Ahlfors regular, and the preceding observation about formulating the BPI condition in terms of cubes permits us to derive the BPI condition for $\overline{Q}$ from that of M.

5.6 Semi-regularity

For our purposes the following small modification of the notion of Minkowski dimension will be the most convenient.

Definition 5.6 (Semi-regular spaces) *Let E be a metric space. Then we say that E is* semi-regular *of dimension d if there is a constant $C > 0$ so that for each subset of E of diameter $R > 0$ and each $0 < r < R$ we can cover the subset by $\leq C\,(R/r)^d$ balls of radius r.*

Note that Ahlfors regularity implies semi-regularity of the same dimension, by Lemma 5.1. Notice also that a subset of a semi-regular space is automatically semi-regular of the same dimension.

If E is semi-regular of dimension d, then E is σ-finite for d-dimensional Hausdorff measure, and so has Hausdorff dimension $\leq d$.

The next definition provides a condition under which we can get an improved bound on the semi-regularity dimension of a subset of an Ahlfors regular space.

Definition 5.7 (Porous sets) *Given a subset E of a metric space M, we say that E is* porous *if there is an $\epsilon > 0$ so that for each $x \in E$ and $0 < r \leq \operatorname{diam} E$ we can find a point $y \in B(x, r)$ such that $\operatorname{dist}(y, E) \geq \epsilon\, r$.*

Note that any subset of a porous set is again porous.

Lemma 5.8 *Let M be a metric space which is Ahlfors regular of dimension d, and suppose that E is a subset of M which is porous. Then E is semi-regular of dimension $d - \eta$, where $\eta > 0$ depends only on the porosity constant for E and the regularity constants for M. In particular E has measure 0 in M.*

The lemma is a simple extension of a well-known fact, and we can extend a standard proof of it to Ahlfors regular metric spaces using cubes. Let Δ_j be a family of cubes for M as before. The porosity condition ensures that if Q is a cube in M, then we can find a subcube Q' of Q with $\operatorname{diam} Q' \geq \alpha \operatorname{diam} Q$ and $Q' \cap E = \varnothing$. Here α depends only on the porosity constant for E and the regularity constants for M, and the existence of central points in cubes (as in (5.10)) is being used here to find Q'.

Let us say this in a slightly different way. If our original cube Q lies in Δ_j, then we are saying that there is a subcube Q' of Q in Δ_{j-N} which does not intersect E, where N is bounded independently of Q.

One can repeat this argument, applying it to the other subcubes of Q in Δ_{j-N}, and then to the subcubes in Δ_{j-2N}, etc., in order to conclude the following. Given a positive integer k, let $\mathcal{Q}_k$ denote the collection of subcubes of Q in Δ_{j-kN} which intersect E. Then

$$\sum_{T \in \mathcal{Q}_k} |T| \leq (1-\beta)^k\, |Q|, \tag{5.12}$$

where β is controlled in terms of ϵ and the regularity constants for M. That is, in each iteration we win some small but fixed proportion, and this is not hard to check by repeating the argument above.

Suppose that we are given a subset A of E of diameter R. Using the properties of cubes one can check that we can cover A by a bounded number of cubes Q with diameters comparable to R. One can then conclude that A can be covered by $\leq C\,(1-\beta)^k\,2^{kNd}$ cubes of diameter 2^{j-kN}, where 2^j is comparable to $\operatorname{diam} E$. From this the lemma follows easily.

Results like Lemma 5.8 (and its converse) have also been observed by Jouni Luukkainen. See [Lu, LS] concerning this and related topics.

6

STRONGER SELF-SIMILARITY FOR BPI SPACES

6.1 The basic result

Roughly speaking, the BPI condition requires that the space have pieces all around that look like each other. There is an annoying ambiguity here; not only might these pieces be small, but a priori they might behave quite differently from place to place. We would be much happier to say that all the pieces look like each other. It turns out that we can always do this, at least to some extent. Let us begin with a couple of definitions.

Definition 6.1 (ACI (almost covers itself)) *Suppose that $(M, d(x,y))$ is a regular metric space of dimension d. We say that M satisfies* ACI *if for each $\delta > 0$ there exist $N, K > 0$ so that for each $x_1, x_2 \in M$ and $0 < r_1, r_2 \leq \operatorname{diam} M$ there exist N closed sets $A_1, \ldots, A_N$ in $\overline{B}(x_1, r_1)$ and N mappings $h_i : A_i \to \overline{B}(x_2, r_2)$ such that $|\overline{B}(x_2, r_2) \backslash \bigcup_{i=1}^{N} h_i(A_i)| \leq \delta\, r_2^d$ and each h_i is K-conformally bilipschitz with scale factor r_2/r_1.*

In other words, the BPI condition requires that there be a substantial piece of $B(x_2, r_2)$ which looks like a subset of $B(x_1, r_1)$, whereas this condition makes the stronger requirement that we can almost cover $B(x_2, r_2)$ by a bounded number of pieces, each of which looks like a piece of $B(x_1, r_1)$. Thus ACI implies BPI by definition, but it turns out that the converse is true.

Proposition 6.2 *BPI implies ACI.*

Before we prove this let us formulate another condition of a similar nature.

Definition 6.3 (EAC (everything almost covers)) *Let M be a regular metric space of dimension d. We say that M satisfies* EAC *if for every $\eta > 0$ there exist $K, N > 0$ with the following property. If $x_1, x_2 \in M$, $0 < r_1, r_2 \leq \operatorname{diam} M$, and $F \subseteq B(x_1, r_1)$ are given, with F closed and $|F| \geq \eta\, r_1^d$, then there exist N closed subsets $F_1, \ldots, F_N$ of F and N mappings $\phi_i : F_i \to B(x_2, r_2)$ such that $|B(x_2, r_2) \backslash \bigcup_{i=1}^{N} \phi_i(F_i)| \leq \eta\, r_2^d$ and each ϕ_i is K-conformally bilipschitz with scale factor r_2/r_1.*

This is different from ACI because it allows us to start with any subset of the first ball that is not too small. All of these conditions can be seen as some kind of mixing conditions, the question is how much mixing do we get. EAC is the nicest kind of mixing, it says that everything gets mixed up. We do not know how to prove that this is true for BPI spaces.

Problem 6.4 *Does BPI imply EAC?*

Note the similarity with concepts from ergodic theory. Remember also our intention of trying to replace mappings (like the ones in dynamical systems and ergodic theory) with pure geometry.

Euclidean spaces do satisfy EAC. This can be proved by iterating Lemma 3.4. (The point is the group structure.) One-dimensional Euclidean spaces are especially nice, because of the rising sun lemma.

Let us now prove the proposition above. The argument is slightly tricky but not a huge surprise, involving fairly standard ingredients of covering and so forth. The rest of this section will be devoted to the proof, and from now on the notation and assumptions in the proposition will be in force.

Lemma 6.5 *Let $(M, d(x,y))$ be a regular metric space of dimension d, and let B be a ball in M with radius $t \leq \operatorname{diam} M$. Let A be a closed subset of $2\overline{B}$ such that $A \cap \overline{B} \neq \varnothing$ and $t^{-d}|\overline{B} \backslash A| \geq \delta > 0$. Then we can find two families of balls $\{B_i\}_{i \in I}$ and $\{\beta_i\}_{i \in I}$ with the following properties:*

$$\text{all the } B_i\text{'s and } \beta_i\text{'s are contained in } 2\overline{B}; \tag{6.1}$$

$$\text{the balls in } \{2B_i\}_{i \in I} \text{ are pairwise disjoint, and similarly for } \{2\beta_i\}_{i \in I}; \tag{6.2}$$

$$2B_i \cap A = \varnothing \text{ for all } i; \tag{6.3}$$

$$\operatorname{radius} B_i = \operatorname{radius} \beta_i \text{ for all } i; \tag{6.4}$$

$$\operatorname{dist}(\beta_i, B_i) \leq 10 \operatorname{radius} B_i; \tag{6.5}$$

$$\text{the } \beta_i\text{'s are centered on } A \text{ and the } B_i\text{'s are centered on } \overline{B}; \tag{6.6}$$

$$\sum_{i \in I} (\operatorname{diam} B_i)^d \geq \epsilon\, t^d. \tag{6.7}$$

Here $\epsilon > 0$ is a constant that depends only on δ and reasonable constants.

We shall sometimes use the phrase "reasonable constants" to refer to constants such as the dimension, the regularity constant for a metric space, the constants that then result for the story of cubes, etc.

The statement of this lemma is pretty technical but its meaning is far from deep. We are assuming that $\overline{B} \backslash A$ is pretty large, as measured by δ, and to prove Proposition 6.2 we shall want to be able to add some things to A in a controlled manner. We are going to make these additions inside $\bigcup_i B_i$. Thus (6.7) says that this union of balls is pretty large, while (6.3) says that it does not interact with A too much. To each B_i there is an associated β_i which does live on A. Thus we shall make constructions in the B_i's that are far from A but which are still associated to A in a certain way.

We shall see these things later, for now let us prove the lemma. We may as well assume that

$$\operatorname{dist}(x, A) \leq \frac{1}{10}\, t \qquad \text{for all } x \in \overline{B}, \tag{6.8}$$

since otherwise the lemma is trivial. (If this assumption fails, we need only one B_i and one β_i, and they are easy to choose.)

We begin with a reasonably trivial covering of $\overline{B}\backslash A$, and we shall reduce it afterwards.

Sublemma 6.6 *There is a family of balls $\{B_j\}_{j\in J}$ in M with the following properties:*

$$B_j = \overline{B}(p_j, r_j), \text{ where } p_j \in \overline{B}\backslash A \text{ and } r_j = \frac{1}{10}\operatorname{dist}(p_j, A); \tag{6.9}$$

the doubles of the B_j's have bounded overlap; (6.10)

$$\overline{B}\backslash A \subseteq \bigcup_{j\in J} B_j. \tag{6.11}$$

This sublemma provides a kind of Whitney decomposition for $\overline{B}\backslash A$ in this context. (See [St1] for Whitney decompositions.) Remember that "bounded overlap" means that no point lies in more than a bounded number of the B_j's, with the bound in this case depending only on reasonable constants (the doubling constant for M is enough).

Let us quickly prove the sublemma. Suppose for the moment that we have a maximal subset X of $\overline{B}\backslash A$ such that

$$d(x,y) > \frac{1}{11}\operatorname{dist}(x, A) \qquad \text{whenever } x, y \in X, x \neq y. \tag{6.12}$$

Then we take $p_j, j \in J$ to be an enumeration of the points in X, and we define B_j as in (6.9). Let us verify (6.10), (6.11). Fix a point $z \in M$ and suppose that $z \in 2B_j$ for some j. Then

$$d(p_j, z) \leq \frac{1}{5}\operatorname{dist}(p_j, A). \tag{6.13}$$

Hence $|\operatorname{dist}(z, A) - \operatorname{dist}(p_j, A)| \leq \frac{1}{5}\operatorname{dist}(p_j, A)$, and therefore

$$\frac{4}{5}\operatorname{dist}(p_j, A) \leq \operatorname{dist}(z, A) \leq \frac{6}{5}\operatorname{dist}(p_j, A). \tag{6.14}$$

Thus if $J(z)$ denotes the set of $j \in J$ such that $z \in 2B_j$, then $d(p_j, z) \leq \frac{1}{4}\operatorname{dist}(z, A)$ for each $j \in J(z)$, while $d(p_j, p_i) > \frac{1}{11}\operatorname{dist}(p_j, A) \geq \frac{5}{66}\operatorname{dist}(z, A)$ when $i, j \in J(z)$, $i \neq j$. The fact that M is doubling imposes a uniform upper bound on the number of elements of $J(z)$. This proves the bounded overlap condition (6.10).

For the covering condition (6.11) we make similar computations. Suppose that $w \in \overline{B}\backslash A$ but $w \notin B_j$ for all j. Then $d(w, p_j) > \frac{1}{10}\operatorname{dist}(p_j, A)$ for all j. Let us check that

$$d(w, p_j) > \frac{1}{11}\operatorname{dist}(w, A). \tag{6.15}$$

We have that

$$d(w, p_j) > \frac{1}{10}\operatorname{dist}(p_j, A) \geq \frac{1}{10}\{\operatorname{dist}(w, A) - d(w, p_j)\}, \tag{6.16}$$

and hence $\frac{11}{10}\, d(w, p_j) > \frac{1}{10}$ dist(w, A). This proves (6.15). We get a contradiction now, because these inequalities imply that we could add w to X and get a set with the same properties, in violation of maximality.

It remains to produce a maximal set X. Given $n > 0$ set $Y_n = \{y \in \overline{B}\backslash A : \operatorname{dist}(y, A) \geq 2^{-n}\}$. These are compact subsets of $\overline{B}\backslash A$, and we have that Y_n increases with n and the union of the Y_n's equals $\overline{B}\backslash A$. Let X_1 denote a maximal subset of Y_1 which satisfies (6.12). This exists, because there is a uniform lower bound on the mutual distances between elements of subsets of Y_1 which satisfy (6.12), and hence there is a uniform upper bound on the number of elements that such a set can have. Thus we simply take X_1 to be a subset of Y_1 which satisfies (6.12) and which has a maximal number of elements. We choose X_n recursively, in such a way that $X_{n-1} \subseteq X_n \subseteq Y_n$, X_n satisfies (6.12), and X_n is maximal with respect to these properties. Again such a maximal set exists because we can get a uniform upper bound for the number of its elements. We take X to be the union of all the X_n's. This set satisfies (6.12), and it is not hard to see that it is a maximal subset of $\overline{B}\backslash A$ with this property.

This completes the proof of Sublemma 6.6. From now on we let $\{B_j\}_{j\in J}$ be as provided there. We conclude from (6.10) and (6.11) that

$$\sum_{j\in J} (\operatorname{diam} B_j)^d \geq C^{-1}\, \delta\, t^d. \tag{6.17}$$

We are also using the regularity of M here, and the constant C depends only on reasonable constants.

Let us apply the Vitali covering lemma to the balls $\{20B_j\}_{j\in J}$ (viewed as a covering of their union). Note that the radii of these balls tend to 0. This gives us a subset I of J such that

$$\text{the balls } 20B_j, j \in I, \text{ are pairwise disjoint,} \tag{6.18}$$

and

$$\bigcup_{j\in J} 20B_j \subseteq \bigcup_{j\in I} 100B_j. \tag{6.19}$$

The family of balls $\{B_j\}_{j\in I}$ is the one that we want. For the β_i's we proceed as follows. For each $i \in I$ we choose $q_i \in A$ such that $\operatorname{dist}(p_i, A) = d(p_i, q_i)$. (We can do this because A is compact, since M is Ahlfors regular.) By our initial assumption (6.8) this quantity is $\leq t/10$, and so $q_i \in \frac{11}{10}\overline{B}$. We take $\beta_i = \overline{B}(q_i, r_i)$, where r_i is as before, in (6.9). It is not hard to check that (6.1)–(6.7) above are satisfied. (For (6.1) we are using (6.8). For (6.2) we are using the observation that $2\beta_i \subseteq 20B_i$.) This proves Lemma 6.5.

Let us now use the lemma to prove Proposition 6.2. Let M be a BPI space, and let $x_1, x_2 . r_1, r_2$ be as in the proposition. Let δ be as in Definition 6.1. Set $B = B(x_2, r_2)$.

Let us say that a set $A \subseteq 2\overline{B}$ is "suitably covered" if there is a finite collection of closed sets $F_l \subseteq \overline{B}(x_1, k\, r_1)$ and mappings $\phi_l : F_l \to M$ such that $A \subseteq \bigcup_l \phi_l(F_l)$

and the ϕ_l's are conformally bilipschitz with bounded constant and scale factor r_2/r_1. We want to find a subset A of $\overline{B}$ which is suitably covered (with uniform bounds on the number of F_l's, the constant k, and the conformal bilipschitz constants) and which satisfies $r_2^{-d}|\overline{B}\backslash A| < \delta$.

We shall produce this set A by repeating a certain procedure a bounded number of times. We begin by applying the BPI condition to get a closed set $A_0 \subseteq \overline{B}$ which is suitably covered (with a single F_l, ϕ_l) and for which $r_2^{-d}\,|A_0|$ is bounded from below.

If $r_2^{-d}\,|\overline{B}\backslash A_0| < \delta$ then we stop. Otherwise we apply Lemma 6.5 with these choices of $\overline{B}$ and δ, and with $A = A_0$. By definition of A_0 we have $F \subseteq \overline{B}(x_1, r_1)$ and $\phi : F \to M$ as above with $\phi(F) = A_0$. Let $\{B_i\}_{i\in I}$ and $\{\beta_i\}_{i\in I}$ be as in Lemma 6.5, with $\beta_i = B(q_i, t_i)$. Remember from Lemma 6.5 that t_i is also the radius of B_i. Set

$$\alpha_i = B\Big(\phi^{-1}(q_i), \eta\,\frac{r_1}{r_2}\,t_i\Big), \tag{6.20}$$

where η is chosen small enough so that the doubles of the α_i's are disjoint. We can do this because of the corresponding condition for the β_i's (6.2) and the conformal bilipschitz condition on ϕ, and we can do this with an η which is bounded from below.

Using the BPI condition we can find a subset of α_i which is a substantial proportion of it in measure and a conformally bilipschitz mapping from this subset into B_i, where the conformal bilipschitz constant is bounded and the scale factor is radius $B_i/$ radius $\alpha_i = \eta^{-1}\,r_2/r_1$. We can combine these mappings to get a mapping ψ that takes a subset of $\bigcup_{i\in I}\alpha_i$ into $\bigcup_{i\in I} B_i$, with the image of ψ being a substantial fraction in measure of $\bigcup_{i\in I} B_i$. (The bound for this fraction does not depend on the choice of η.) This mapping is also conformally bilipschitz with bounded constant and scale factor r_2/r_1. (This constant does involve the choice of η.) This fact comes from the conformal bilipschitz condition of the pieces together with the conformal bilipschitz condition on ϕ. That is, suppose that we take two points in the domain of ψ and we want to estimate the distance between their images. If the two points come from the same α_i then we simply use the estimate for that piece. (η is involved here.) Suppose now that the two points u and v come from α_i and α_j with $i \neq j$. We reduce to the conformal bilipschitz condition for ϕ as follows. We start with the fact that the doubles of α_i and α_j are disjoint, to conclude that $d(u, v)$ is comparable to the distance between the centers of α_i and α_j, which is $d(\phi^{-1}(q_i), \phi^{-1}(q_j))$. Thus

$$d(u, v) \approx d(\phi^{-1}(q_i), \phi^{-1}(q_j)). \tag{6.21}$$

On the other hand the doubles of the B_i's are also disjoint, by (6.2), and this implies that the distance between $\psi(u)$ and $\psi(v)$ is comparable to the distance between the centers of B_i and B_j. We also have that the distance between the centers of B_i and B_j is comparable to the distance between the centers of β_i and β_j. This is not hard to check, using Lemma 6.5. The upper bound uses (6.4), (6.5), and for the lower bound one can use also (6.2). The bottom line is that the distance between $\psi(u)$ and $\psi(v)$ is comparable to the distance between the centers of β_i and β_j, which is $d(q_i, q_j)$. That is,

$$d(\psi(u), \psi(v)) \approx d(q_i, q_j). \tag{6.22}$$

Thus we can derive the conformal bilipschitz condition for ψ in this case from that of ϕ.

Let A_1 be the closure of the union of A_0 and the image of ψ. The good news is that $|A_1|$ is substantially larger than $|A_0|$. Indeed, the image of ψ is disjoint from A_0, and so

$$|A_1 \backslash A_0| \geq |\{ \text{ the image of } \psi\}| \geq C^{-1} \sum_{i \in I} (\operatorname{diam} B_i)^d \geq C^{-1} \epsilon \, r_2^d. \tag{6.23}$$

The middle inequality comes from the construction of ψ, with C depending on the BPI constants of M, and the last inequality comes from (6.7), with ϵ controlled as in Lemma 6.5. (Note that C does *not* depend on the choice of η above.) Of course A_1 is "suitably covered" in the sense above, using the suitable covering of A_0 and the closure of the domain of ψ. Note that the domain of ψ is contained in $B(x_1, 2\, r_1)$, say, but this is allowed.

If $r_2^{-d} \, |\overline{B} \backslash A_1| < \delta$ then we stop. Otherwise we can produce a larger set $A_2 \subseteq \overline{B}$ which is suitably covered. The construction is slightly messier now, because the number of new conformally bilipschitz mappings that we have to construct is equal to the number used so far, i.e., 2. Nonetheless the construction is basically the same. We apply Lemma 6.5 to get collections of balls $\{B_i\}_{i \in I}$ and $\{\beta_i\}_{i \in I}$ associated to $A = A_1$, and we define the α_i's in the same manner as before. Again we use the BPI condition to build some nice mappings into $\bigcup_{i \in I} B_i$. Now there is a complication that we have to distinguish between the i's for which the center of β_i lies on A_0 or on $A_1 \backslash A_0$. We build two conformally bilipschitz mappings, one defined on a subset of the union of the α_i's which corresponds to the first set of i's, the second mapping for the second set of α_i's. Again we get the right bounds, and the union of the images of these two mappings covers a substantial fraction of $\bigcup_{i \in I} B_i$. We take A_2 to be the closure of the union of A_1 and the images of these two mappings, and we continue to make progress as in (6.23).

Similarly, if $r_2^{-d} \, |\overline{B} \backslash A_2| < \delta$ then we stop, otherwise we continue with the construction to produce a set A_3 which is suitably covered and substantially larger than A_2.

It is very important here that the constants in (6.23) and its descendants do not depend on the level of the construction. They come from the BPI constants and Lemma 6.5. As long as we are missing a subset of $\overline{B}$ of mass $\geq \delta r_2^d$ we can continue to make progress.

(The analogues of η in the later stages of the construction do deteriorate. That is, they remain bounded from below, but with a bound that depends on the conformal bilipschitz constants of the mappings at the previous generation, and these constants can increase with the level of the construction. This is the price we pay to keep the constants in (6.23) held fixed.)

The domains of the conformally Lipschitz mappings whose images cover A_k are contained in $B(x_1, 2^k r_1)$. That is, we have to keep increasing the radius on the domain side. The conformal bilipschitz constants also increase, and faster than that. The main point though is that with each stage of the construction $r_2^{-d}|A_k|$ is increasing by a definite amount each time, with a uniform lower bound. This means that the construction has to end in a finite number of steps – i.e., $r_2^{-d}|\overline{B}\backslash A_k| < \delta$ in a finite number of steps – and that we have a bound on the number of steps required. This leads to a uniform bound on the conformal bilipschitz constant, and on the amount of expansion (2^k) of the ball $B(x_1, r_1)$ that we have to permit in order to get our suitable coverings.

This proves Proposition 6.2, modulo the fact that r_1 has to be relabelled to accommodate the expansion that we have to allow. (This is not at all a problem, we simply have to start with a smaller radius. The constants do not depend on the radii.)

6.2 A corollary

Corollary 6.7 *Suppose that $(M, d(x,y))$ is a BPI metric space of dimension d. Then for every $\epsilon > 0$ there is a $K > 0$ so that if $x_1, x_2 \in M$ and $0 < r_1, r_2 \leq \operatorname{diam} M$, and if $Z \subseteq B(x_1, r_1)$ is measurable and satisfies $|Z| \geq \epsilon r_1^d$, then there is a measurable subset Z_0 of Z with $|Z_0| \geq \epsilon K^{-1} r_1^d$ and a K-conformally bilipschitz mapping $\phi : Z_0 \to B(x_2, r_2)$ with scale factor r_2/r_1.*

That is, we can strengthen the BPI condition by requiring that the domain of the conformally bilipschitz mapping lie in a prescribed set whose measure is not too small. This follows from Proposition 6.2, which says that we can require instead that the image lies in a prescribed set which is not to small. That is, Corollary 6.7 follows from Proposition 6.2 by reversing the roles of domain and image, of (x_1, r_1) with (x_2, r_2). Notice, however, that we do not know whether we can prescribe supersets for both the domain and range simultaneously. This amounts to Problem 6.4.

6.3 Regular subsets

Proposition 6.8 (Regular subsets are BPI) *If $(M, d(x,y))$ is a BPI metric space of dimension d, then any subset of M which is Ahlfors regular of dimension d is a BPI space in its own right.*

To prove this we shall prove that BPI spaces satisfy the following weaker version of the EAC property, in which the initial set is itself regular.

Proposition 6.9 (Regular subsets almost cover) *Let $(M, d(x,y))$ be a BPI metric space of dimension d, and let F be an Ahlfors regular subset of M of dimension d. Let $x_1 \in F$, $x_2 \in M$, $0 < r_1 \leq \operatorname{diam} F$, and $0 < r_2 \leq \operatorname{diam} M$ be given, and let $\delta > 0$ be given also. Then there exist closed subsets $F_1, \ldots, F_N$*

of $F \cap B(x_1, r_1)$ and mappings $\phi_i : F_i \to B(x_2, r_2)$ such that each ϕ_i is K-conformally bilipschitz with scale factor r_2/r_1 and

$$\left| B(x_2, r_2) \setminus \bigcup_{i=1}^{N} \phi_i(F_i) \right| \le \delta\, r_2^d, \tag{6.24}$$

where N and K do not depend on the choices of (x_1, r_1) or (x_2, r_2), and depend on F only through its regularity constant.

Proposition 6.8 will follow easily once we prove this.

Let us now prove Proposition 6.9. The argument is nearly the same as for Proposition 6.2. Let everything be given as in the proposition. Set $B = B(x_2, r_2)$.

We say that a closed set $A \subseteq 2\overline{B}$ is "nicely covered" if there is a finite collection of closed subsets G_i of $F \cap B(x_1, k\, r_1)$ and mappings $\psi_i : G_i \to M$ such that $A \subseteq \bigcup_i \psi_i(G_i)$ and each ψ_i is conformally bilipschitz with scale factor r_2/r_1. We require also that the conformal bilipschitz constants, the expansion factor k, and the number of sets G_i be uniformly bounded (by constants that do not depend on (x_1, r_1) or (x_2, r_2)). We want to show that there is an A which is nicely covered such that $|\overline{B} \setminus A| \le \delta\, r_2^d$.

From Corollary 6.7 we know that there is a nicely covered set $A_0 \subseteq \overline{B}$ with $|A_0| \ge C^{-1}\, r_2^d$. This constant C is permitted to depend on the regularity constant for F, as are the other constants in the proof.

If $|B(x_2, r_2) \setminus A_0| \le \delta\, r_2^d$ then we are happy and we stop. Otherwise we apply Lemma 6.5 with these choices of B and δ and with $A = A_0$. We use the same argument as in the proof of Proposition 6.2 to build a new set $A_1 \subset 2\overline{B}$ which is nicely covered and whose measure is larger than that of A_0 by a definite amount. In repeating this argument we must make some small changes; if $\{\alpha_i\}_{i \in I}$ is the collection of balls that we defined before, then we have to work with $\alpha_i \cap F$ rather than the α_i's themselves for building the new mappings. That is, the new mappings that we construct must be defined on subsets of $\alpha_i \cap F$ instead of just α_i. This is not a problem, because Corollary 6.7 provides us with the necessary substitute for the BPI condition that we need to do this. We are also using here the fact that we have a good lower bound for $|\alpha_i \cap F|$, because of the assumption that F is regular.

With this small modification the rest of the argument is exactly the same as it was before. This proves Proposition 6.9.

6.4 Some remarks about EAC

We do not know whether BPI spaces must satisfy the EAC condition, but there is a weaker version that is easy to check. We first state the condition as a separate definition.

Definition 6.10 (CPI) *Let $(M, d(x, y))$ be a metric space which is σ-finite with respect to its Hausdorff measure H^d, $0 < d < \infty$. We say that M satisfies CPI ("covered by (countably many) pieces of itself") if for each H^d-measurable set*

$A \subseteq M$ with $H^d(A) > 0$ there is a sequence of measurable subsets $\{A_i\}$ of A and bilipschitz mappings $\phi_i : A_i \to M$ such that $H^d(M \backslash \bigcup_i \phi_i(A_i)) = 0$.

This should be compared with the concept of rectifiability (Definition 3.1). Rectifiable sets satisfy CPI. This is not hard to prove, using points of density, as in the argument below.

Proposition 6.11 *BPI spaces satisfy CPI. In fact, for each Borel set $A \subseteq M$ of positive measure there is a sequence $\{A_i\}$ of Borel subsets of A and a sequence of bilipschitz mappings $\phi_i : A_i \to M$ such that $M \backslash \bigcup_i \phi_i(A_i)$ has measure 0, and we may choose the ϕ_i's so that they have uniformly bounded conformal Lipschitz constants.*

This is pretty easy to prove. Let M be a BPI space of dimension d, and let A be a subset of M of positive measure, which we can even take to be Borel measurable. Then almost all points of A are points of density, as in (5.4), and so we can find a ball B such that $|B|^{-1}\,|B \backslash A|$ is as small as we like. Let $\{(x_i, r_i)\}$ be a countable dense subset of $M \times (0, \operatorname{diam} M)$. Since M has BPI there are Borel sets $A_i \subseteq A$ and conformally bilipschitz mappings $\phi_i : A_i \to B(x_i, r_i)$ with bounded constant such that $|\phi_i(A_i)| \geq \epsilon\, r_i^d$ for some $\epsilon > 0$ that does not depend on i. The set $M \backslash \bigcup_i \phi_i(A_i)$ must have measure 0 because it cannot have any points of density. This proves the proposition.

7

BPI EQUIVALENCE

7.1 Basic facts

We want to prove some basic facts about BPI equivalence (Definition 1.6). We begin with the following, which amounts to saying that it does not take too much to get BPI equivalence in the presence of the BPI condition for the spaces.

Proposition 7.1 *Let $(M, d(x, y))$ and $(N, \rho(u, v))$ be two BPI spaces with the same dimension d. Then M and N are BPI equivalent if and only if there is a subset A of M with positive measure which is bilipschitz equivalent to a subset of N.*

Of course the "only if" part is automatic, it is the "if" part that needs a proof.

Without loss in generality A is closed. Since A has a point of density there is a ball $B = B_M$ in M such that

$$|B|^{-1}\,|B\backslash A| \tag{7.1}$$

is as small as we like. (We shall continue to follow here our convention of writing $|E|$ for the measure of E, but now this depends on the space in which E lies.) Let R denote the radius of B, and let ψ denote the bilipschitz embedding of A into N.

Let $x \in M$, $0 < r \leq \operatorname{diam} M$, $u \in N$, $0 < t \leq \operatorname{diam} N$ be given. By definition of BPI spaces we can find a closed set $E \subseteq B_M(x, r)$ with $|E| \geq \epsilon\, r^d$ and a K-conformally bilipschitz mapping $\phi : E \to B$ with scale factor R/r, where ϵ and K are the BPI constants for M. By replacing ϵ with $\epsilon/2$ we may assume that $\phi(E) \subseteq A$ (by demanding that (7.1) be small enough, depending on K and ϵ).

Set $F = \psi(\phi(E))$. Thus F is contained in a ball β with radius equal to $\operatorname{diam} F$ and with $|F| \geq C^{-1}\,|\beta|$, where C depends on the preceding constants and the bilipschitz constant for ϕ. Using Corollary 6.7 we get a closed subset G of F with $|G| \geq \gamma\,|F|$ and an L-conformally bilipschitz mapping $\theta : G \to B_N(u, t)$ with scale factor $t/\operatorname{diam} F$. Altogether $\theta \circ \psi \circ \phi : F \to B_N(u, t)$ is a conformally bilipschitz mapping with bounded constant and scale factor t/r, and this is what we need to conclude BPI equivalence. This proves the proposition.

Remark 7.2 (About bounds) The notion of BPI equivalence is a quantitative one, i.e., it comes with bounds. The proof of Proposition 7.1 provides bounds which depend on the bilipschitz constant for ϕ but not on the measure of A. (Actually the bounds depend only on the conformal bilipschitz constant for ϕ.)

Proposition 7.3 (Unions) *Let $(M, d(x,y))$ be a metric space, and suppose that M_1 and M_2 are two subspaces of M which are BPI with the same dimension d. Then M_1 and M_2 are BPI equivalent if and only if their union is BPI, with bounds in both directions.*

To be honest, notice that the union of two Ahlfors regular sets of the same dimension is Ahlfors regular, but that the constant is bad if the distance between them is large compared to the maximum of their diameters.

The "only if" part follows directly from the definitions. For the "if" part we may as well assume that neither M_1 nor M_2 is contained in the other, because of Proposition 6.8. Thus we can find balls B_1, B_2 in M such that B_i is centered on M_i and disjoint from the other M_j. We may even choose the balls to have the same radius. The BPI assumption now provides a bilipschitz mapping from a subset of $B_1 \cap M_1$ into $B_2 \cap M_2$, and BPI equivalence now follows from Proposition 7.1.

Proposition 7.4 *Let $(M, d(x,y))$ and $(N, \rho(u,v))$ be two BPI spaces with the same dimension d.*

(a) *If M and N are BPI equivalent, then there is a sequence of closed sets $\{A_i\}$ in M and bilipschitz mappings $\phi_i : A_i \to N$ such that $N \setminus \bigcup_i \phi_i(A_i)$ has measure 0. The ϕ_i's may be chosen to have uniformly bounded conformal bilipschitz constants; the A_i's may be chosen among subsets of any given Borel subset of M of positive measure.*

(b) *If M and N are not BPI equivalent, and if $A \subseteq M$ and $\phi : A \to N$ is bilipschitz, then $H^d(A) = H^d(\phi(A)) = 0$.*

Part (a) follows easily from successive applications of Proposition 6.11. Part (b) is an immediate consequence of Proposition 7.1. We shall derive stronger results in Section 9.6.

Proposition 7.5 *BPI equivalence is an equivalence relation on BPI sets.*

Only transitivity is not immediately clear from the definition, and that follows from Propositions 7.4 and 7.1. Note that we get appropriate bounds. (See Remark 7.2.)

7.2 BPI equivalence and uniform rectifiability

Proposition 7.6 *Let $(M, d(x,y))$ be a metric space which is Ahlfors regular of dimension d. Then M is uniformly rectifiable if and only if M is BPI and BPI equivalent to $\mathbf{R}^d$.*

We are cheating slightly here, since uniform rectifiability was officially defined (in Definition 3.2) only for subsets of Euclidean spaces, but we can extend it to Ahlfors regular metric spaces in the obvious manner.

The fact that uniformly rectifiable spaces are BPI was proved before, in Proposition 3.3. That they are BPI equivalent to Euclidean spaces follows immediately from the definition. Similarly, BPI equivalence to a Euclidean space implies uniform rectifiability by the definitions.

We like to think of an equivalence class of BPI spaces as defining a geometry, with Euclidean geometry being an example. We can also define the notion of "uniform rectifiability with respect to a BPI equivalence class", but we do not get anything new this way, beyond membership in the given equivalence class. We shall state and prove this now.

Definition 7.7 (Uniform rectifiability with respect to BPI spaces) *Let $(M, d(x,y))$ and $(N, \rho(u,v))$ be two Ahlfors regular metric spaces of dimension d, and suppose that M is BPI. We say that N* is uniformly rectifiable with respect to M *if there exist constants $K, \theta > 0$ so that for each $u \in N$, $0 < t \leq \operatorname{diam} N$ there exist $x \in M$, $0 < r \leq \operatorname{diam} M$, a closed set $F \subseteq B_M(x,r)$ with $|F| \geq \theta\, r^d$, and a K-conformally bilipschitz mapping $\phi : F \to B_N(u,t)$ with scale factor t/r.*

This condition clearly holds if N is BPI and BPI equivalent to M. The converse is also true, as in the next result.

Proposition 7.8 *If $(M, d(x,y))$ and $(N, \rho(u,v))$ are two Ahlfors regular metric spaces of dimension d, with M BPI and N uniformly rectifiable with respect to M. Then N is BPI and BPI equivalent to N.*

If N is a subset of M, then the uniform rectifiability of N with respect to M is automatic, but the BPI conditions are not. In this special case the conclusions of the proposition are contained already in Propositions 6.8 and 7.1.

We shall derive Proposition 7.8 from a generalization of Propositions 6.2 and 6.9. We state this generalization in gory detail in part because it provides a more complete version of some of our earlier constructions.

Proposition 7.9 *Let $(M, d(x,y))$ and $(N, \rho(u,v))$ be two Ahlfors regular metric spaces of dimension d. Suppose that there are constants $K, \theta > 0$ so that for all $x \in M$, $u \in N$, and $0 < r \leq \min(\operatorname{diam} M, \operatorname{diam} N)$ there is a closed set $F \subseteq B_M(x,r)$ with $|F| \geq \theta\, r^d$ and a K-bilipschitz mapping $\phi : F \to B_N(u,r)$. Then for every $\delta > 0$ there exist $L, p > 0$ so that for each $x \in M$, $u \in N$, and $0 < r \leq \min(\operatorname{diam} M, \operatorname{diam} N)$ there are closed sets $F_1, \ldots, F_p \subseteq \overline{B}_M(x,r)$ and L-bilipschitz mappings $\phi_i : F_i \to \overline{B}_N(u,r)$ such that $|\overline{B}_N(u,r) \backslash \bigcup_{i=1}^{p} \phi_i(F_i)| \leq \delta\, r^d$.*

This is another variation on the theme of one space almost covering another. The proof of this is very similar to the proof of Proposition 6.2. The proof would be almost exactly the same if we allowed the radius of the ball in N to be independent of the radius of the ball in M, as opposed to requiring that they be equal as we do here. This requirement leads to a couple of complications that we shall address in a moment. We shall not really need this extra requirement here, but it is nice to record it. Before we do that let us explain how Proposition 7.9 implies Proposition 7.8.

Let M and N be Ahlfors regular metric spaces of dimension d such that M is BPI and N is uniformly rectifiable with respect to M. Notice that Corollary 6.7 implies that N satisfies a stronger version of the uniform rectifiability condition in which we get to choose the point $x \in M$ and the radius $0 < r \leq \operatorname{diam} M$ instead of saying only that they exist. This is practically the same as BPI equivalence, we are missing only the BPI condition for N. We can apply now Proposition 7.9 with the roles of M and N reversed to conclude that any given ball in M is almost covered by bilipschitz images of pieces of a ball in N. For this we may have to dilate N to accommodate the case where $\operatorname{diam} M > \operatorname{diam} N$, and thereby break the equality of the radii of the balls in M and N, but this does not really matter because we do not need to know the radius of the ball in N for the next step, so long as we control the conformal bilipschitz constants and the scale factors. The next step is to use the uniform rectifiability of N with respect to M (strengthened as above to permit us to choose the point and radius in M) to obtain the BPI condition for N. This proves Proposition 7.8 assuming Proposition 7.9.

The logic of the preceding argument could be simplified if we knew that M satisfies EAC. Our extra contortions circumvent this ignorance.

Now let us prove Proposition 7.9, using the argument of Proposition 6.2. We hardly used the stronger assumption of BPI before, nor did we really need to be working with only one space, rather than the two that we have now. There are a couple of complications now, though, stemming from the fact that we ask for a correspondence between balls in M and N only when they have the same radius, and not for all pairs of radii as before.

The first complication arises in the construction of the mapping ψ in the proof of Proposition 6.2. In making this construction we used the fact that there is a conformally bilipschitz mapping from a substantial subset of α_i onto a substantial subset of B_i. In the present setting this is dangerous because the radius of α_i will normally be much smaller than the radius of B_i. This comes from the η in the definition (6.20) of α_i. The ratio r_1/r_2 in (6.20) plays no role in the present context, because we are taking our balls in domain and range to have the same radius r. The η is more dangerous because it depends on the (conformal) bilipschitz constant of the mapping ϕ that we have in our hands at the moment of (6.20). As the construction proceeds from stage to stage these bilipschitz constants deteriorate, and the η's which are used deteriorate also. This causes a problem when we want to produce a bilipschitz mapping from a substantial part of α_i onto a substantial fraction of B_i. It did not cause a problem before, when we had the BPI condition which permitted us to compare balls of arbitrary radii, whereas here we are supposed to restrict ourselves to balls of the same radii. To resolve this problem we simply construct more bilipschitz mappings. In the present situation we keep the same choices of η and the α_i's, and we observe that we can find a bounded number of bilipschitz mappings from subsets of α_i into B_i such that the union of their images is a substantial fraction of B_i, with bounds on their bilipschitz constants. This follows from the fact that we can cover B_i with a bounded number of balls of the same radius as α_i. This bound depends on η, but we do not mind.

Thus when we make the mapping ψ at the first stage of the construction in the proof of Proposition 6.2, or the several mappings like ψ in later stages of the construction, we might now have to make many more mappings to accommodate the shrinking η. That is, we have to do this to ensure that we are always getting a substantial proportion of B_i, so that we can get an estimate like (6.23). As the construction proceeds we have to use more and more mappings, because the constant η depends on the bilipschitz constants from the preceding stage of the construction, which are permitted to deteriorate. However, we can still maintain an estimate like (6.23) to ensure that we are making a definite amount of progress at each stage. This implies also that the construction will stop in a bounded number of stages. Thus we still get bounds, even if they are much worse than before.

This is the first complication. The second is that the argument in the proof of Proposition 6.2 gives some spreading in the domains of our bilipschitz mappings. That is, if we follow the argument of Proposition 6.2, then we get sets F_i and mappings ϕ_i as above, except that the F_i's live in $B_M(x, L\,r)$ where L depends on δ but not x, u, or r. This was not a problem before, because we could change our radii as we liked, but now it is an issue. We can resolve it with a fairly general argument. We can cover $B_N(u, r)$ with a collection $\mathcal{B}$ of $\leq C\,L^d$ balls of radius $L^{-1}\,r$. We apply the argument just described (from the proof of Proposition 6.2) to cover a large *proportion* of each ball in $\mathcal{B}$ – meaning all of the ball except for a set of measure $\delta(L^{-1}\,r)^d$ – by a bounded number of bilipschitz images of subsets of $B_M(x, L \cdot (L^{-1}\,r)) = B_M(x, r)$. That is, we apply the previous argument with $B_N(u, r)$ replaced by the given element of $\mathcal{B}$, and with $B_M(x, r)$ replaced by $B_M(x, (L^{-1}\,r))$. Note that we are really using the same δ as before, only the radii are being changed. (If this were not the case we would be in trouble, we need to know the size of the expansion L in advance.) By combining these coverings we get a covering of a large proportion of $B_N(u, r)$ by bilipschitz images of subsets of $B_M(x, r)$. Now "large proportion" means except for a subset of measure $\leq C\,\delta\,r^d$. This constant C comes from the covering and does not depend on δ, not even through L. In other words it does not bother us, and the proof is complete. (We pay for the cost of making the covering by again increasing the number of bilipschitz mappings that we need, and this increase depends on δ through L, but we do not mind.)

These complications would not occur if we permitted the radii of the balls in M and N in Proposition 6.2 to be independent of each other, and this special case would suffice for proving Proposition 7.8.

Notice that Proposition 7.9 implies Proposition 6.2 (one has only to rescale the metric). We can also derive Proposition 6.9 from Proposition 7.9 (as opposed to proving it by repeating the argument of Proposition 6.2, as we did before). The point is that by allowing ourselves to work with two different sets we can apply Proposition 7.9 to the situation of Proposition 6.9, with the metric space N in Proposition 7.9 taken to be the space M from Proposition 6.9, and with the metric space M in Proposition 7.9 taken to be the regular subset in Proposition 6.9. One has to permit a rescaling of the metric to accommodate the different radii of the balls in Proposition 6.9, and one has to use Corollary 6.7 to get that the hypotheses of Proposition 7.9 hold.

7.3 A strengthening of BPI equivalence

Proposition 7.10 (Finding bilipschitz-equivalent regular subsets) *Let $(M, d_M(x,y))$ and $(N, d_N(u,v))$ be two BPI spaces of dimension d which are BPI equivalent. Let $x \in M$, $0 < r \leq \operatorname{diam} M$, $u \in N$, and $0 < t \leq \operatorname{diam} N$ be given. Then we can find a compact regular set $F \subseteq \overline{B}_M(x,r)$ (of dimension d) with $\operatorname{diam} F \geq \eta r$ and a K-conformally bilipschitz mapping $\phi : F \to B_N(u,t)$ with scale factor t/r, where η, K, and the regularity constant for F depend only on the BPI constants and the constants for the BPI equivalence between M and N (and on the dimension and regularity constants, but not on the choices of x, r, u, or t).*

The difference between this result and the statement of BPI equivalence is that we are requiring F to be Ahlfors regular too. This result is new even in the context of uniformly rectifiable sets (with respect to Euclidean spaces).

Corollary 7.11 *Let $(M, d_M(x,y))$ and $(N, d_N(u,v))$ be two BPI spaces of dimension d which are BPI equivalent. Given $x \in M$ and $u \in N$ we can find a regular subset G of M which contains x and a K-conformally bilipschitz mapping $\psi : G \to N$ such that $\psi(x) = u$. We may choose G and ψ so that $\eta r \leq \operatorname{diam} G \leq r$ and ψ has scale factor t/r, where r and t are finite positive real numbers such that $r \leq \operatorname{diam} M$ and $t \leq \operatorname{diam} N$. If M and N are both unbounded then we can choose G to be unbounded and ψ to have scale factor 1. The constants K and η and the regularity constant for G depend only on the BPI constants and the constants for the BPI equivalence between M and N (and on the dimension and regularity constants).*

The corollary is an easy consequence of the proposition. The main point is to use Lemma 5.4 to produce sequences of balls in M and N which converge to x and u, respectively, with radii decreasing in a geometric progression, with their distances to x and u decreasing at the same rate (to within a bounded factor), but with the radii somewhat smaller than the distances, so that the balls have disjoint doubles. One makes the balls correspond in the obvious manner, and then the proposition can be used to find regular sets in the balls in M and bilipschitz embeddings of these regular subsets into the balls in N, with suitable bounds. These regular sets in the balls in M can be combined with the point x itself to produce a regular subset of M as in the corollary. (Approximately the same issue arises also in the next section, in a slightly trickier way that we shall treat more explicitly.)

Let us now prove the proposition. The idea is to start with a bilipschitz embedding on a substantial set as in the definition of BPI equivalence and to augment it using again BPI equivalence. The proof takes some space, and will occupy us for the rest of the section (except for some remarks at the end about extensions of the method).

Let everything be given as in the proposition. Let $F_0 \subseteq B_M(x,r)$ be a closed set with $|F_0| \geq \epsilon\, r^d$ such that there is a conformally bilipschitz embedding $\phi_0 : F_0 \to B_N(u,t)$ with scale factor t/r and bounded conformal bilipschitz constant. Without loss of generality we may assume that $r = t = 1$, so that ϕ_0 is bilipschitz with bounded constant. (We could always simply rescale the metrics.)

Let $\theta > 0$ be small, to be chosen later. θ will be our threshold for small densities. We certainly require that θ be much smaller than $|F_0|$. Let j_0 be the smallest positive integer such that

$$D(j_0) = \{y \in F_0 : |B_M(y, 2^{-j_0}) \cap F_0| < \theta\, 2^{-j_0 d}\} \tag{7.2}$$

is nonempty. If there is no such positive integer, then F_0 is already regular, and we are finished, and so we may assume that j_0 exists. Let $E(j_0)$ denote a maximal subset of $D(j_0)$ such that $d_M(y,z) \geq 2^{-j_0+1}$ whenever $y, z \in E(j_0)$ and $y \neq z$. Thus

$$D(j_0) \subseteq \bigcup_{y \in E(j_0)} B_M(y, 2^{-j_0+1}), \tag{7.3}$$

by maximality, and the balls $B_M(y, 2^{-j_0})$, $y \in E(j_0)$, are disjoint.

Let $\rho \in (0, 1/10)$ be another small constant, to be specified later. (ρ will be chosen before θ, so that θ is allowed to depend on ρ.) For each $y \in E(j_0)$ let $A(y, j_0)$ be a closed subset of $\overline{B}_M(y, \rho\, 2^{-j_0})$ such that $|A(y, j_0)| \geq \epsilon\, (\rho\, 2^{-j_0})^d$ and such that there is an L-bilipschitz mapping $\psi_{y,j_0} : A(y, j_0) \to B_N(\phi_0(y), \rho\, 2^{-j_0})$, where the constants ϵ and L come from BPI equivalence. Set

$$F_{j_0} = \Big(F_0 \backslash \bigcup_{y \in E(j_0)} B_M(y, 2^{-j_0})\Big) \cup \Big(\bigcup_{y \in E(j_0)} A(y, j_0)\Big). \tag{7.4}$$

It is easy to see that F_{j_0} is closed. We shall use it to replace F_0. The idea is that we look for the places where F_0 is too thin, we cut them out, and we plug new sets into the holes which are thicker than F_0 was.

Define $\phi_{j_0} : F_{j_0} \to N$ by

$$\phi_{j_0} = \phi_0 \quad \text{on} \quad F_0 \backslash \bigcup_{y \in E(j_0)} B_M(y, 2^{-j_0})$$

$$\phi_{j_0} = \psi_{y,j_0} \quad \text{on} \quad A(y, j_0),\ y \in E(j_0).$$

It is not hard to show that ϕ_{j_0} is bilipschitz when ρ is small enough, but we shall not worry about estimates until later, after finishing the construction.

Let us explain now how to construct sets F_j and mappings $\phi_j : F_j \to N$ for all $j \geq j_0$. Suppose that F_{j-1} and ϕ_{j-1} have been chosen, and we want to define F_j and ϕ_j. Set

$$D(j) = \{y \in F_{j-1} : |B_M(y, 2^{-j}) \cap F_{j-1}| < \theta\, 2^{-jd}\}. \tag{7.5}$$

If $D(j) = \varnothing$ then we are reasonably happy and we set $F_j = F_{j-1}, \phi_j = \phi_{j-1}$. Otherwise we take a maximal subset $E(j)$ of $D(j)$ such that $d_M(y,z) \geq 2^{-j+1}$ whenever $y, z \in E(j)$, $y \neq z$. Again we have that

$$D(j) \subseteq \bigcup_{y \in E(j)} B_M(y, 2^{-j+1}), \tag{7.6}$$

by maximality, and that the balls $B_M(y, 2^{-j})$, $y \in E(j)$, are pairwise disjoint.

Notice that

$$E(j) \subseteq F(j-1) \qquad \text{when } n > j_0, \quad E(j_0) \subseteq F_0, \tag{7.7}$$

and that

$$|B_M(y, 2^{-j}) \cap F_{j-1}| < \theta\, 2^{-jd} \qquad \text{when } y \in E(j). \tag{7.8}$$

We now use the BPI equivalence of M and N in the same manner as before. We obtain for each $y \in E(j)$ a closed set $A(y,j)$ with

$$A(y,j) \subseteq \overline{B}_M(y, \rho\, 2^{-j}) \qquad \text{and} \qquad |A(y,j)| \geq \epsilon\, (\rho\, 2^{-j})^d \tag{7.9}$$

and an L-bilipschitz mapping

$$\psi_{y,j} : A(y,j) \to B_N(\phi_{j-1}(y), \rho\, 2^{-j}). \tag{7.10}$$

We define F_j and $\phi_j : F_j \to N$ by

$$F_j = \Big(F_{j-1} \setminus \bigcup_{y \in E(j)} B_M(y, 2^{-j})\Big) \cup \Big(\bigcup_{y \in E(j)} A(y,j)\Big) \tag{7.11}$$

and

$$\begin{aligned} \phi_j &= \phi_{j-1} \quad \text{on} \quad F_j \setminus \bigcup_{y \in E(j)} B_M(y, 2^{-j}) \\ \phi_j &= \psi_{y,j} \quad \text{on} \quad A(y,j),\ y \in E(j) \end{aligned} \tag{7.12}$$

as before. Notice again that F_j is closed.

By repeating this process we obtain F_j and ϕ_j for all $j \geq j_0$.

Let us come now to estimates. Fix $j \geq j_0$ and $y \in E(j)$ for the moment. We want to see first how the pieces can interact with each other. We have that

$$\begin{aligned} &F_k \cap B_M(y, 2^{-j}) \\ &\quad \subseteq B_M(y, \rho\, 2^{-j+1}) \cup \{B_M(y, 2^{-j}) \setminus B(y, (1-\rho) 2^{-j})\} \end{aligned} \tag{7.13}$$

for all $k \geq j$. This follows from (7.11) and (7.9) by summing the appropriate geometric series. More precisely, (7.11) and (7.9) imply that

$$\operatorname{dist}(x, F_{l-1}) \leq \rho\, 2^{-l} \qquad \text{whenever } x \in F_l. \tag{7.14}$$

Using this and (7.11) we get (7.13) by summing a geometric series.

Similarly,

$$\phi_k(F_k \cap B_M(y, 2^{-j-1})) \subseteq B_N(\phi_{j-1}(y), \rho\, 2^{-j+1}) \tag{7.15}$$

for all $k \geq j$. (The $\phi_{j-1}(y)$ on the right-hand side should be replaced by $\phi_0(y)$ when $j = j_0$.) When $k = j$ this is immediate from (7.12), but with the better radius $\rho\, 2^{-j}$ on the right-hand side. For $k > j$ we observe that if $x \in F_k \cap B_M(y, 2^{-j-1})$, then

$$\operatorname{dist}(\phi_k(x), \phi_{k-1}(F_{k-1} \cap B_M(y, 2^{-j-1}))) \leq \rho\, 2^{-k}, \tag{7.16}$$

because of (7.12) and (7.10) (with j replaced by k). (We are also using (7.13) here to know that x lies in $F_{k-1} \cap B_M(y, 2^{-j-1})$ instead of $F_{k-1} \backslash B_M(y, 2^{-j-1})$.) Once we have this estimate it is easy to derive (7.15).

Before proceeding let us set some notation for the combinatorics. Given $p, q \in M$ and $j \geq j_0$, define $R_j(p, q)$ by

$$\begin{aligned} R_j(p,q) &= 1 \quad \text{if } p, q \in B_M(y, 2^{-j-1}) \\ &\qquad \text{for some } y \in E(j), \\ R_j(p,q) &= 0 \quad \text{otherwise.} \end{aligned} \tag{7.17}$$

Thus $R_j(p, q) = 1$ when p and q are "related" at the scale of 2^{-j}, and $R_j(p, q) = 0$ when they are unrelated.

Let us now get bilipschitz estimates for the ϕ_j's. Fix $j \geq j_0$ and $p, q \in F_j$. The idea is that at each stage when we make modifications to our previous definitions, these modifications are sufficiently small and localized so as not to disturb what we did before.

Suppose first that $R_j(p, q) = 1$. Then there exists $y \in E(j)$ such that $p, q \in B_M(y, 2^{-j-1})$. By definition of F_j we have that $p, q \in A(y, j)$ and $\phi_j(p) = \psi_{y,j}(p)$, $\phi_j(q) = \psi_{y,j}(q)$. Then

$$L^{-1}\, d_M(p, q) \leq d_N(\phi_j(p), \phi_j(q)) \leq L\, d_M(p, q), \tag{7.18}$$

since we know that $\psi_{y,j}$ is L-bilipschitz.

Now suppose that $R_j(p, q) = 0$ but $R_l(p, q) = 1$ for some $j_0 \leq l < j$, and let l be the largest such integer. Thus there is a $y \in E(l)$ such that $p, q \in B_M(y, 2^{-l-1})$. Let $k(p)$ denote the smallest integer k, $l < k \leq j$, such that $R_k(p, p) = 1$. (If no such k exists, set $k(p) = \infty$.) By definition $k(p)$ marks the first time after the lth stage of the construction and before the end of the jth stage of the construction that p is potentially disturbed by the construction. Define $k(q)$ similarly. In particular, if $k(p) = k(q) = \infty$, then $\phi_j(p) = \phi_l(p) = \psi_{y,l}(p)$ and $\phi_j(q) = \phi_l(q) = \psi_{y,l}(q)$, and we have (7.18) again. Now suppose that $k(p), k(q) < \infty$. In this case there exist $z \in E(k(p))$ and $w \in E(k(q))$ such that $p \in B_M(z, 2^{-k(p)-1})$, $q \in B_M(w, 2^{-k(q)-1})$. From (7.13) we actually have that

$$p \in B_M(z, \rho\, 2^{-k(p)+1}), \qquad q \in B_M(w, \rho\, 2^{-k(q)+1}). \tag{7.19}$$

Claim 7.12 $z, w \in A(y, l)$.

Let us check that $z \in A(y, l)$, the argument is the same for w. Notice first that $z \in F_{k(p)-1}$, because of (7.7). Next observe that

$$R_m(z, z) = 0 \qquad \text{when } l < m < k(p). \tag{7.20}$$

Indeed, if this were not the case, then we would have $z \in B_M(a, 2^{-m-1})$ for some $a \in E(m)$, $l < m < k(p)$. From (7.13) we would actually have that $z \in B_M(a, \rho\, 2^{-m+1})$. Combining this with (7.19) we get that

$$p \in B_M(a, \rho(2^{-m+1} + 2^{-k(p)+1})). \tag{7.21}$$

Since $\rho(2^{-m+1} + 2^{-k(p)+1}) < 2^{-m-1}$ (remember that we chose $\rho < 1/10$), this contradicts the minimality of $k(p)$. Thus we have (7.20). Since $z \in F_{k(p)-1}$ we conclude that $z \in F_l$.

On the other hand $p \in B_M(y, 2^{-l-1})$ and $d_M(p, z) < 2^{-k(p)-1}$, and so $z \in B_M(y, \frac{3}{2} 2^{-l-1})$. Again (7.13) applies to give $z \in B_M(y, 2^{-l-1})$. Since $z \in F_l$ we conclude that $z \in A(y, l)$ from the definition (7.11), (7.4) of $F(l)$. This proves the claim.

We want to estimate $d_N(\phi_j(p), \phi_j(q))$. From (7.15) we have that

$$\phi_j(F_j \cap B_M(z, 2^{-k(p)-1})) \subseteq B_N(\phi_{k(p)-1}(z), \rho\, 2^{-k(p)+1}). \tag{7.22}$$

This means that

$$d_N(\phi_j(p), \phi_{k(p)-1}(z)) \leq \rho\, 2^{-k(p)+1}. \tag{7.23}$$

On the other hand (7.20) implies that $\phi_{k(p)-1}(z) = \phi_l(z)$, and Claim 7.12 implies that $\phi_l(z) = \psi_{y,l}(z)$. Thus $\phi_{k(p)-1}(z) = \psi_{y,l}(z)$. Therefore

$$d_N(\phi_j(p), \psi_{y,l}(z)) \leq \rho\, 2^{-k(p)+1}. \tag{7.24}$$

Similarly

$$d_N(\phi_j(q), \psi_{y,l}(w)) \leq \rho\, 2^{-k(q)+1}. \tag{7.25}$$

From Claim 7.12 and the bilipschitz condition on $\psi_{y,l}$ we have that

$$L^{-1}\, d_M(z, w) \leq d_N(\psi_{y,l}(z), \psi_{y,l}(w)) \leq L\, d_M(z, w). \tag{7.26}$$

Of course we want to convert this into an estimate for p and q.

Let us check that

$$\frac{1}{2}\, d_M(z, w) \leq d_M(p, q) \leq 2\, d_M(z, w). \tag{7.27}$$

We know that $R_{k(p)}(p, q) = 0$ (by the maximality of l), and this implies that $q \notin B_M(z, 2^{-k(p)-1})$. Using (7.13) we get that $q \notin B_M(z, (1 - \rho)2^{-k(p)})$, and hence

$$d_M(p,q) \geq 2^{-k(p)-1} \tag{7.28}$$

because of (7.19). Similarly $d_M(p,q) \geq 2^{-k(q)-1}$, and hence

$$d_M(p,q) \geq 2^{-k(p)-2} + 2^{-k(q)-2}. \tag{7.29}$$

This together with (7.19) yields (7.27), if ρ is small enough, and we also get that

$$d_M(z,w) \geq 2^{-k(p)-3} + 2^{-k(q)-3}. \tag{7.30}$$

Combining these various estimates we can conclude that

$$(CL)^{-1}\, d_M(p,q) \leq d_N(\phi_j(p), \phi_j(q)) \leq CL\, d_M(p,q). \tag{7.31}$$

That is, we start with (7.26), and we use our estimates to make suitable corrections. The main point is that we need $\rho\, L$ to be small.

This last is of course the kind of estimate that we wanted. We want to have a similar estimate also when one of $k(p), k(q)$ is finite and the other is infinite. This case is a mixture of the two preceding ones, and we omit the details. (The main point is that if $k(p) = \infty$, then $\phi_j(p) = \phi_l(p) = \psi_{y,j}(p)$, and we do nothing for p, while we repeat the same kind of argument above to deal with q, $k(q) < \infty$, i.e., we choose w as before and we make the same estimates as before.)

We are left now with the case where $R_i(p,q) = 0$ for all i, $j_0 \leq i \leq j$. This case is handled in essentially the same manner as before, but with l replaced by 0. That is, there is no l really, no y, and no $B_M(y, 2^{-j-1})$. Now F_0 plays the role that $A(y,l)$ did before, and ϕ_0 plays the role of both ϕ_l and $\psi_{y,l}$. One still defines $k(p)$ and $k(q)$ as before, except that one restricts k to the range $j_0 \leq k \leq j$ instead of $l < k \leq j$. One considers separately the cases where $k(p), k(q)$ are finite or infinite, as before. With these substitutions the arguments works as before, with only cosmetic changes.

Let us summarize this discussion of bilipschitz estimates in a lemma.

Lemma 7.13 (Bilipschitz estimates) *If ρL is sufficiently small (less than an absolute constant), then*

$$(CL)^{-1}\, d_M(p,q) \leq d_N(\phi_j(p), \phi_j(q)) \leq CL\, d_M(p,q), \tag{7.32}$$

for all $p, q \in F_j$ and $j \geq j_0$, where C is an absolute constant.

Next we need to estimate the densities of the F_j's from below. The constant ρ should be considered as chosen now, as in the lemma, and our lower bounds for the densities are permitted to depend on ρ.

Lemma 7.14 *Let $j \geq j_0$ be given. If θ is small enough (depending on ρ and ϵ), then for each $x \in F_{j-1}$ and $r > 0$ we have that*

$$H^d(F_j \cap B_M(x, r + 2^{-j+1})) \geq H^d(F_{j-1} \cap B_M(x,r)). \tag{7.33}$$

(If $j = j_0$, then we replace F_{j-1} on the right-hand side with F_0.)

That is, we do not lose mass in most situations. This follows easily from the construction of F_j from F_{j-1}. Suppose that $B_M(x,r)$ touches any of the balls that were removed in the construction (7.11) of F_j. Then $B_M(x, r+2^{-j+1})$ contains all of the $A(y,j)$'s that correspond to the balls that were removed. The $A(y,j)$'s are pairwise disjoint by construction, and they each have mass $\geq \epsilon\,(\rho\, 2^{-j})^d$ (as in (7.9)). If θ is small enough, then we win more than we lost, as in (7.8). This proves the lemma.

Lemma 7.15 *If θ is small enough (depending on ρ and ϵ), then for every $j \geq j_0$ and $z \in F_j$ we have that*

$$H^d(F_j \cap B_M(z, 2^{-j+5})) \geq \theta\, 2^{-jd}. \tag{7.34}$$

This inequality also holds for $0 < j < j_0$ and $z \in F_0$, with F_j replaced by F_0.

This is also an easy consequence of our construction. Suppose first that $0 < j < j_0$, so that $F_j = F_0$. Then (7.34) follows from the definition of j_0, given just before (7.2). (That is, j_0 was chosen to be the smallest positive integer such that $D(j_0)$ is not empty.) Thus we assume that $j \geq j_0$. Let $z \in F_j$ be given.

Assume first that

$$z \in F_{j-1} \backslash D(j) \tag{7.35}$$

and

$$F_{j-1} \cap B_M(z, 2^{-j}) \subseteq F_j, \tag{7.36}$$

with the convention that F_{j-1} is replaced with F_0 when $j = j_0$. Then

$$H^d(F_j \cap B_M(z, 2^{-j})) \geq \theta\, 2^{-jd}, \tag{7.37}$$

because of the definitions of j_0, $D(j_0)$, and $D(j)$. (See (7.2) and (7.5).) In this case we have our desired lower bound. If (7.35) fails, so that either $z \in F_j \backslash F_{j-1}$ or $z \in D(j)$, then there is a point $y \in E(j)$ such that $d_M(z,y) \leq 2^{-j+1}$. This follows from (7.4), (7.11), (7.3), and (7.6). Similarly, if (7.35) holds but (7.36) fails, then there is a point $y \in E(j)$ such that $d_M(z,y) \leq 2^{-j+1}$, because of (7.4) and (7.11). In either case there is $y \in E(j)$ with $d_M(z,y) \leq 2^{-j+1}$, and we have that $A(y,j) \subseteq B_M(z, 2^{-j+5})$. This implies (7.34), because of the lower bound (7.9) on the mass of the $A(y,j)$'s. This proves the lemma.

From now on we assume that θ is chosen small enough, as in the preceding pair of lemmas. We shall not impose any further conditions on θ.

Lemma 7.16 *The Hausdorff distance between F_j and F_{j-1} is $\leq 2^{-j+1}$, i.e.,*

$$\sup_{x \in F_{j-1}} \operatorname{dist}(x, F_j) \leq 2^{-j+1} \quad \text{and} \quad \sup_{x \in F_j} \operatorname{dist}(x, F_{j-1}) \leq 2^{-j+1}. \tag{7.38}$$

When $j = j_0$ we replace F_{j-1} with F_0.

This is easy to derive from the definition (7.11) of F_j.

Lemma 7.17 *Suppose that $k \geq j \geq j_0$, $x \in F_k$, and $r > 0$. Then*

$$H^d(F_k \cap B_M(x, r + 2^{-j+2})) \geq H^d(F_j \cap B_M(x_j, r)), \tag{7.39}$$

for some $x_j \in F_j$. This also holds when $j < j_0$ if we replace F_j with F_0.

Indeed, let k, j, x, r be given as in the lemma. Set $x_k = x$, and choose $x_l \in F_l$ recursively for $l = k-1, k-2, \ldots, j$ so that $d_M(x_{l-1}, x_l) \leq 2^{-l+1}$, using Lemma 7.16. (If $l < j_0$, then $x_l \in F_0$, and we take $x_{l-1} = x_l$.) Set $r_l = r + \sum_{i=j+1}^{l} 2^{-i+2}$, $r_j = r$. We get that

$$\begin{aligned} H^d(F_l \cap B_M(x_l, r_l)) &\geq H^d(F_l \cap B_M(x_{l-1}, r_l - 2^{-l+1})) \\ &\geq H^d(F_{l-1} \cap B_M(x_{l-1}, r_l - 2^{-l+1} - 2^{-l+1})) \\ &= H^d(F_{l-1} \cap B_M(x_{l-1}, r_{l-1})) \end{aligned} \tag{7.40}$$

for all $l > j$, using the triangle inequality and then Lemma 7.14. By repeating this estimate we get that

$$H^d(F_k \cap B_M(x_k, r_k)) \geq H^d(F_j \cap B_M(x_j, r_j)). \tag{7.41}$$

This proves the lemma, since $x_k = x$, $r_j = r$, and $r_k \leq r + 2^{-j+2}$.

Lemma 7.18 *Let j and k be positive integers with $k > \max(j, j_0)$, and let $x \in F_k$ be given. Then*

$$H^d(F_k \cap B_M(x_k, 2^{-j+5} + 2^{-j+2})) \geq \theta\, 2^{-jd}. \tag{7.42}$$

Indeed, we can apply Lemma 7.17 to conclude that

$$H^d(F_k \cap B_M(x_k, 2^{-j+5} + 2^{-j+2})) \geq H^d(F_j \cap B_M(x_j, 2^{-j+5})), \tag{7.43}$$

where $x_j \in F_j$, and where we take $F_j = F_0$ when $j < j_0$. The lemma follows by applying Lemma 7.15.

We are now ready to finish the proof of Proposition 7.10. Lemma 7.16 implies that the F_j's converge in the Hausdorff metric to a nonempty compact subset F of M. (The notion of Hausdorff convergence is reviewed in Section 8.1 below.) It is not hard to check that F is regular, as follows. The upper bounds on the mass of balls centered on F are inherited from the regularity of M. For the lower bounds we want to know that $r^{-d}\, H^d(F \cap \overline{B}_M(x, r))$ is not too small when $x \in F$ and $0 < r < 1$. Suppose to the contrary that $r^{-d}\, H^d(F \cap \overline{B}_M(x, r))$ is very small for some choice of x and r. This means that we can cover $F \cap \overline{B}_M(x, r)$ by a family of sets V_i such that $\sum_i (\operatorname{diam}_M V_i)^d$ is very small compared to r^d. We may as well assume that the V_i's are open, we can always make them a little larger without disturbing the sum too much. Since $F \cap \overline{B}_M(x, r)$ is compact we may assume that there are only finitely many of the V_i's. By Hausdorff convergence we conclude that $F_k \cap \overline{B}_M(x, r/2)$ is also covered by the V_i's when k is large enough. This covering implies then that $H^d(F_k \cap \overline{B}_M(x, r/2))$ is small compared to r^d when k is very large, because of the Ahlfors regularity of M. This cannot happen because of Lemma 7.18. Thus we conclude that F is regular. Because we get a lower bound on the mass of balls of radius up to 1 we also conclude that there is a uniform lower bound on the diameter of F.

So we have our regular set F now, and we need a bilipschitz embedding ϕ. Morally we take the limit of the ϕ_j's, but strictly speaking this is not legal because the domains of the ϕ_j's are not increasing or anything like that. This is not a serious matter, however. Let us first verify the following convergence property of the ϕ_j's: given $j \geq j_0$, $x \in F_j$, and $z \in F_{j+1}$, we have that

$$d_N(\phi_j(x), \phi_{j+1}(z)) \leq C\,(2^{-j} + d_M(x, z)). \tag{7.44}$$

This is not hard to establish, using the definition (7.12) of ϕ_j, (7.10), and the fact that the Lipschitz constants for the ϕ_j's are bounded, as in Lemma 7.13. (C is allowed to depend on L, but not on x, z, or j.) To define ϕ now we argue as follows. Given $x \in F$ and $j \geq j_0$ we can find $x_j \in F_j$ so that $d_M(x, x_j) \leq 2^{-j+2}$, by Lemma 7.16. We set

$$\phi(x) = \lim_{j \to \infty} \phi_j(x_j). \tag{7.45}$$

This limit exists because of (7.44). One can see for the same reason that it does not depend on the choice of the approximating sequence $\{x_j\}$, but we do not really need that. The main point is that the bilipschitz bounds in Lemma 7.13 extend to ϕ, because of the convergence.

This completes the proof of Proposition 7.10, except for the minor point that F may not lie inside $B_M(x, 1)$ and $\phi(F)$ may not lie in $B_N(u, 1)$. It is not hard to show that $F \subseteq B_M(x, 2)$ and $\phi(F) \subseteq B_N(u, 2)$, however. Indeed, we had $F_0 \subseteq B_M(x, 1)$ and $\phi_0(F_0) \subseteq B_N(u, 1)$ at the beginning, and at the jth stage of the construction we did not move further than $\rho\, 2^{-j}$ from where we were in either the domain or the range. See (7.11), (7.12), (7.9), and (7.10). Summing the geometric series we obtain that domain and range are contained in balls of radius 2, and this is good enough for the proposition, since we could always adjust the initial choices of the radii.

The proof of Proposition 7.10 is now finished.

Remark 7.19 We never really needed to assume that M and N have the same dimension in the proof above. We do not even need N to be regular either, for that matter. It is enough to assume that M is regular, that N is complete, and that for each $x \in M$, $0 < r \leq \operatorname{diam} M$, $u \in N$, and $0 < t \leq \operatorname{diam} N$ there is a closed set $A \subseteq B_M(x, r)$ with $|A| \geq \epsilon\, r^d$ and a L-conformally bilipschitz mapping $\phi : A \to B_N(u, t)$ with scale factor t/r. Under these assumptions we get the same result as before, and with the same proof.

Also, we may restrict ourselves to $r = t$ (equal radii in domain and range) in the hypotheses so long as we do the same in the conclusion. Again the proof does not change. (See (7.9) and (7.10).)

7.4 Mappings with big bilipschitz pieces

There is a special case of Proposition 7.10 which is rather amusing. Before we state it we need an auxiliary definition.

Definition 7.20 (BBP (big bilipschitz pieces)) *Let $(M, d_M(x, y))$ be a regular metric space of dimension d, and let $(N, d_N(u, v))$ be another metric space. Let $f : M \to N$ be a Lipschitz mapping. We say that f has* BBP *if there exist $\epsilon, L > 0$ such that for each $x \in M$ and $0 < r \leq \operatorname{diam} M$ there is a closed set $A \subseteq B_M(x, r)$ with $|A| \geq \epsilon\, r^d$ such that the restriction of f to A is L-bilipschitz.*

See [D3, J2, DS3] for some results pertaining to the BBP condition for mappings in the context of Euclidean geometry. (The concept of "weakly bilipschitz" from [DS3] is also relevant here.)

Proposition 7.21 (Bilipschitz on regular subsets) *Let $(M, d_M(x, y))$ be a regular metric space of dimension d, let $(N, d_N(u, v))$ be a complete metric space, and let $f : M \to N$ be a Lipschitz mapping with BBP. Then for each $x \in M$ and $0 < r \leq \operatorname{diam} M$ there is a compact regular set $F \subseteq B_M(x, r)$ (of dimension d) such that $\operatorname{diam} F \geq \eta t$ and such that the restriction of f to F is bilipschitz. The bilipschitz constant for this restriction, the regularity constant for F, and the constant η depend only on the Lipschitz and BBP constants for f and the regularity constants for M.*

Proposition 7.21 is proved in almost exactly the same manner as Proposition 7.10 was. The point is simply that all the mappings ϕ and ψ which appeared in the proof (with subscripts) can be taken to be restrictions of f. To understand this it is helpful to make the reduction to the case where f is 1-Lipschitz. We can certainly take the initial mapping ϕ_0 in the proof of Proposition 7.10 to be the restriction of f to a set F_0 like the one before. The only other place where we introduce new mappings is in (7.10), and we can take restrictions of f for them as well. (It is here that the reduction to the case where f is 1-Lipschitz helps to make the proof look exactly like the previous one. Note that if we are at the point of wanting to define $\psi_{y,j}$, then we have that $\phi_{j-1}(y) = f(y)$, which permits us to take $\psi_{y,j}$ to be a piece of f again.)

Note that in Proposition 7.10 we assumed that the radius in the image was no greater than the diameter of the image. In the present case we do not need to make this assumption. Indeed it is implicit in the BBP condition modulo a constant factor, because BBP implies that

$$\operatorname{diam} N \geq C^{-1} \operatorname{diam} M \tag{7.46}$$

for a suitable constant C.

This proves Proposition 7.21.

Corollary 7.22 *Same assumptions as above. Given $x \in M$ and $0 < r \leq \operatorname{diam} M$ we can find a regular subset G of M such that the restriction of f to G is bilipschitz, $x \in G$, and $\eta r \leq \operatorname{diam} G \leq r$. If M is unbounded we can choose G to be unbounded. We have the same kind of bounds as before.*

This follows from the proposition in much the same way that Corollary 7.11 was derived from Proposition 7.10. There is a minor complication, however. Let M, N, f, and x be as above. We want to take a nice sequence of balls in M, apply the proposition to get certain regular subsets of these balls, and then combine these sets to get the regular set described in the corollary. The problem is that we have to be careful to make sure that the images do not get too close to each other.

Using Lemma 5.4 we can find a sequence of points $\{x_j\}_{j=-\infty}^{m}$ in M such that

$$K^{m+1} \geq \operatorname{diam} M \tag{7.47}$$

and

$$K^j \leq d_M(x_j, x) \leq K^{j+1} \tag{7.48}$$

for all j, where $K \geq 10$ depends only on the regularity constants for M. We allow $m = \infty$ to accommodate the case when $\operatorname{diam} M = \infty$. Using the BBP property we can choose points $y_j \in M$ such that

$$d_M(y_j, x_j) \leq \frac{1}{10} K^j \qquad \text{and} \tag{7.49}$$

$$C^{-1} K^j \leq d_N(f(y_j), f(x)) \leq C\, K^j, \tag{7.50}$$

where C depends on the BBP constants for f. (The upper bound follows from the Lipschitz condition on f, but for the lower bound use the BBP to say that $\operatorname{diam}_N f(B_N(x_j, \frac{1}{10} K^j)) \geq C^{-1} K^j$.) By taking the y_j's and skipping terms we can get a sequence $\{z_k\}_{k=-\infty}^{n}$ such that

$$T^{n+1} \geq \operatorname{diam} M \tag{7.51}$$

$$T^k \leq d_M(z_k, x) \leq T^{k+1} \tag{7.52}$$

$$T^{k-\frac{1}{3}} \leq d_N(f(z_k), f(x)) \leq T^{k+\frac{1}{3}} \tag{7.53}$$

for a large constant $T \geq 100$ that depends only on the regularity and BBP constants. Again we allow $n = \infty$ when $\operatorname{diam} M = \infty$. The last estimate implies that

$$d_N(f(z_k), f(z_l)) \geq \frac{1}{4} T^{k-\frac{1}{3}} \tag{7.54}$$

when $l < k$.

Set $B_k = B_M(z_k, a\, T^{k-\frac{1}{3}})$, where $a > 0$ is a small constant, chosen small enough (depending on the Lipschitz constant of f) so that

$$\operatorname{dist}_N(f(B_k), f(B_l)) \geq \frac{1}{2} T^{k-\frac{1}{3}} \tag{7.55}$$

when $l < k$. Of course we also have that the doubles of the B_k's are disjoint so long as we do not make stupid choices. To prove Corollary 7.22 we apply Proposition 7.21 to each of these balls B_k to get a sequence of regular sets $\{F_k\}_{k=-\infty}^{n}$, and then we take $G = (\bigcup_{k=-\infty}^{n} F_k) \cup \{x\}$. It is easy to check that G is regular, using the regularity of the F_k's, and that the restriction of f to G is bilipschitz, using the uniform bilipschitzness of the restrictions of f to the F_k's and the bounds above. This proves the corollary.

8

CONVERGENCE OF METRIC SPACES

One of the basic tools for studying BPI spaces is to take limits of them, and to pass to weak tangents. This will enable us to make compactness arguments, which will sometimes permit us to derive quantitative statements from qualitative statements. In this chapter we discuss convergence of metric spaces in general, leaving BPI spaces to the next chapter. We discuss also convergence of mappings between spaces that are converging, convergence of subsets, and convergence of measures on converging spaces.

8.1 Hausdorff convergence

Let $(K, \rho(x,y))$ be a compact metric space, and let $\mathcal{H}$ denote the collection of nonempty compact subsets of K. We can define a metric $D(\cdot,\cdot)$ on $\mathcal{H}$ by setting

$$D(E,F) = \sup_{x \in E} \operatorname{dist}(x,F) + \sup_{y \in F} \operatorname{dist}(y,E) \tag{8.1}$$

for all $E, F \in \mathcal{H}$. It is a well-known exercise that this indeed defines a metric.

The second basic fact is that $(\mathcal{H}, D(\cdot,\cdot))$ is actually a compact metric space (when K is compact). This is another well-known exercise. One reasonable approach is to show that $\mathcal{H}$ is totally bounded, which means that for each $\epsilon > 0$ there is a finite covering of $\mathcal{H}$ by balls of radius ϵ, and then to show that $\mathcal{H}$ is complete. It is very easy to show that $\mathcal{H}$ is totally bounded using the corresponding property for K, and it is not terribly difficult to show completeness either.

In short, given a sequence of nonempty compact subsets of a compact metric space, we can find a subsequence which converges to another nonempty compact set, with convergence as above.

8.2 Convergence in Euclidean spaces

What about convergence of subsets of Euclidean spaces, where we no longer have compactness? For any locally compact space one can pass to the one-point compactification, and that amounts to what we shall do here, but we prefer to formulate this in slightly different language.

Definition 8.1 (Convergence of subsets of Euclidean spaces) *Let $\{F_j\}$ be a sequence of nonempty closed subsets of $\mathbf{R}^n$, and let F be another nonempty closed subset of $\mathbf{R}^n$. We say that $\{F_j\}$* converges to *F if*

$$\lim_{j \to \infty} \sup_{x \in F_j \cap B(0,R)} \operatorname{dist}(x,F) = 0 \tag{8.2}$$

and

$$\lim_{j\to\infty} \sup_{y\in F\cap B(0,R)} \operatorname{dist}(y, F_j) = 0 \tag{8.3}$$

for all $R > 0$. These suprema are interpreted to vanish when the relevant sets of competitors $F_j \cap B(0,R)$, $F \cap B(0,R)$ are empty.

Lemma 8.2 (Existence of limits of sets) *Let $\{F_j\}$ be a sequence of closed subsets of $\mathbf{R}^n$, with $F_j \neq \varnothing$ for all j, and suppose that there exists an $r > 0$ such that $F_j \cap B(0,r) \neq \varnothing$ for all j. Then there is a subsequence of $\{F_j\}$ which converges to a nonempty closed subset F of $\mathbf{R}^n$ in the sense defined above.*

This can be proved in the same manner as for Hausdorff convergence in the preceding section, or one can reduce to the case of convergence of a compact space by adding a point at infinity to convert $\mathbf{R}^n$ into a sphere. In this case convergence in the sphere corresponds exactly to the convergence above, at least if the sets do not go running off to infinity, which is prevented by our assumptions. (Alternatively, one could define a notion of "convergence to infinity".)

Remark 8.3 (Convergence of products) Notice that this notion of convergence behaves well under products. That is, if $\{F_j\}$ is a sequence of nonempty closed subsets of $\mathbf{R}^n$ which converges to F in the sense above, and if $\{G_j\}$ is a sequence of nonempty closed subsets in $\mathbf{R}^m$ which converges to the nonempty set G, then $F_j \times G_j$ converges to $F \times G$. This is easy to check.

8.3 Convergence of mappings

Definition 8.4 (Convergence of mappings) *Let $\{F_j\}$ be a sequence of closed subsets of $\mathbf{R}^n$, which are nonempty for all j, and let F be another nonempty closed subset of $\mathbf{R}^n$. Assume that $\{F_j\}$ converges to F in the sense of Definition 8.1. Let $(N, d_N(u,v))$ be a metric space, and let $\phi_j : F_j \to N$ and $\phi : F \to N$ be mappings. We say that $\{\phi_j\}$* converges to ϕ *if for each sequence $\{x_j\}$ in $\mathbf{R}^n$ with $x_j \in F_j$ for all j which converges to some point $x \in F$ we have that*

$$\lim_{j\to\infty} \phi_j(x_j) = \phi(x). \tag{8.4}$$

It is easy to see that this limit is unique if it exists, because the convergence assumption on the F_j's ensures that for each $x \in F$ there is a sequence $\{x_j\}$ in $\mathbf{R}^n$ with $x_j \in F_j$ for all j which converges to x.

In order to have an existence result for limits of mappings we need an equicontinuity assumption.

Definition 8.5 (Equicontinuity and uniform boundedness) *Let $\{F_j\}$ be a sequence of nonempty closed subsets of $\mathbf{R}^n$, and let $f_j : F_j \in \mathbf{R}^n$ be a sequence of mappings on them. We say that the f_j's are* equicontinuous on bounded sets *if for each bounded set B in $\mathbf{R}^n$ and each $\epsilon > 0$ we can find a $\delta > 0$ such that*

$$|f_j(x) - f_j(y)| < \epsilon \qquad \text{whenever } x, y \in F_j \text{ and } |x-y| < \delta. \tag{8.5}$$

(That is, this should hold for all j.) We say that the f_j's are uniformly bounded on bounded subsets *if for each bounded subset B of $\mathbf{R}^n$ we have that*

$$\sup_j \sup_{x \in B \cap F_j} |f_j(x)| < \infty. \tag{8.6}$$

Lemma 8.6 (Existence of limits of mappings) *Let $\{F_j\}$ be a sequence of nonempty closed subsets of $\mathbf{R}^n$, let F be another nonempty closed subset of $\mathbf{R}^n$, and assume that $\{F_j\}$ converges to F in the sense of Definition 8.1. Let $(N, d_N(u,v))$ be a metric space for which closed and bounded subsets are compact. Let $\phi_j : F_j \to N$ be a sequence of mappings which are equicontinuous on bounded sets and uniformly bounded on bounded subsets of the F_j's. Then there is a subsequence of the F_j's and ϕ_j's for which we have convergence in the sense of Definition 8.4 above. If the ϕ_j's are all L-Lipschitz, or L-bilipschitz for some L, then the limiting mapping enjoys the same property.*

This is pretty straightforward. Let everything be given as in the lemma. Let E be a countable dense subset of F, and for each $x \in E$ choose a sequence $\{x_j\}$ in $\mathbf{R}^n$ with $x_j \in F_j$ for all j which converges to x. Then $\{\phi_j(x_j)\}$ is a bounded sequence in N, by our assumptions, and hence has a convergent subsequence. The usual diagonalization argument permits us to choose a subsequence of the F_j's and ϕ_j's for which $\phi_j(x_j)$ converges for each choice of $x \in E$. (Initially we are asking for convergence only along the countably many sequences $\{x_j\}$ that we have chosen, and not all sequences.) This is the subsequence we shall use. Once we have convergence in this special way it is easy to get convergence in the sense of Definition 8.1, using equicontinuity. (This is very much like the standard proofs of the Arzela–Ascoli theorem.) The fact that the Lipschitz and bilipschitz conditions pass to the limit is a straightforward exercise. This proves the lemma.

Let us record another useful fact.

Lemma 8.7 (Uniform convergence of mappings) *Suppose that $\{F_j\}$ is a sequence of nonempty closed subsets of $\mathbf{R}^n$ which converges to another nonempty closed set F in $\mathbf{R}^n$ in the sense of Definition 8.1. Let $(N, d_N(u,v))$ be a metric space, and let $\phi_j : F_j \to N$ and $\phi : F \to N$ be mappings such that $\{\phi_j\}$ converges to ϕ in the sense of Definition 8.4. Assume that the ϕ_j's are equicontinuous on bounded subsets. Then we have uniform convergence on bounded sets, in the sense that for each bounded set B in $\mathbf{R}^n$ and $\epsilon > 0$ there exist $\delta, K > 0$ such that*

$$d_N(\phi_j(x), \phi(y)) < \epsilon \tag{8.7}$$

whenever $x \in F_j$, $j > K$, $y \in F$, and $|x - y| < \delta$.

This is not hard to check, using standard compactness arguments.

8.4 Convergence of spaces

What if we want to talk about convergence of abstract metric spaces? One has to make assumptions, of course, a kind of uniform compactness. There is a theory for doing this, which is described in [BS, Gr1, Gr2, Pe]. For the sake of simplicity we shall avoid this theory in a inelegant way, using embeddings into Euclidean spaces.

In dealing with unbounded metric spaces there are some problems which stem from the possibility of things escaping off to infinity. We shall avoid this by working with "pointed" metric spaces, in which a basepoint has been chosen. In applications this normalization will frequently be natural anyway. Occasionally it is a minor nuisance, and in any case it is not a deep issue.

Definition 8.8 (Pointed metric spaces) *A* pointed metric space *is a triple* $(M, d(x,y), p)$*, where* $(M, d(x,y))$ *is a metric space and* p *is a distinguished point called the basepoint.*

Of course we can continue to speak of Ahlfors regularity, the doubling property, BPI, etc., in the context of pointed metric spaces. For these concepts the basepoint plays no special role.

Now we want to define convergence for pointed metric spaces. We shall restrict ourselves to spaces which are doubling – the only kind that we use here – so that Assouad's embedding theorem permits us to reduce to the case of subsets of $\mathbf{R}^n$. We shall not have isometric embeddings, and so we shall have to impose an additional condition to ensure the convergence of the metrics. We shall need to show also that this notion of convergence does not depend on the embeddings in a too-serious way. The definition below is slightly messy, but it suits our needs technically, for the purpose of existence results, for instance. We shall give a more usable formulation afterwards.

Note that one can get rid of the doubling condition using more general embeddings (embeddings with uniform bounds on their moduli of continuity and the moduli of continuity of their inverses), but we shall not pursue this point. The doubling condition is convenient and it is typically present in our applications.

Definition 8.9 (Convergence of metric spaces) *Let* $\{(M_j, d_j(\cdot,\cdot), p_j)\}$ *be a sequence of pointed metric spaces, and let* $(M, d(\cdot,\cdot), p)$ *be another pointed metric space. We say that* M_j *converges to* M *if the following conditions obtain. We ask first for an* $\alpha \in (0,1]$ *and* K*-bilipschitz embeddings* $f_j : (M_j, d_j(\cdot,\cdot)^\alpha) \to (\mathbf{R}^n, |x-y|)$ *and* $f : (M, d(\cdot,\cdot)^\alpha) \to (\mathbf{R}^n, |x-y|)$ *with* $f_j(p_j) = 0$ *for all* j *and* $f(p) = 0$*. Here* α*,* L*, and* n *are permitted to be arbitrary, but they should not depend on* j*. We require that* $f_j(M_j)$ *converges to* $f(M)$ *in the sense of Definition 8.1. (This means implicitly that the sets* $f_j(M_j)$*,* $f(M)$ *should be closed, which amounts to the requirement that the* M_j*'s and* M *be complete metric spaces. Note that our embeddings force the* M_j*'s and* M *to be doubling with uniformly bounded constants.)*

In addition we require that the real-valued functions $d_j(f_j^{-1}(x), f_j^{-1}(y))$ *on* $f_j(M_j) \times f_j(M_j)$ *converge to* $d(f^{-1}(x), f^{-1}(y))$ *on* $f(M) \times f(M)$ *in the sense of Definition 8.4. (Remark 8.3 is relevant here.)*

Remark 8.10 (Uniform convergence of the metrics) Lemma 8.7 can always be applied to the functions $d_j(f_j^{-1}(x), f_j^{-1}(y))$ on $f_j(M_j) \times f_j(M_j)$ in the situation of Definition 8.9 . That is, these functions are always equicontinuous on bounded sets. Indeed, because of the bilipschitz property of the f_j's this assertion reduces to the fact that if $(N, \rho(\cdot,\cdot))$ is any metric space, then $\rho(x,y)$ is a Lipschitz function on $N \times N$, with a Lipschitz constant that depends only on the way that one chooses to put a metric on $N \times N$. (If one uses the sum metric $D((u_1,u_2),(v_1,v_2)) = \rho(u_1,v_1) + \rho(u_2,v_2)$, then the Lipschitz constant can be taken to be 1.) Notice incidentally that the functions $d_j(f_j^{-1}(x), f_j^{-1}(y))$ on $f_j(M_j) \times f_j(M_j)$ are also uniformly bounded on bounded subsets under the conditions of the definition.

Lemma 8.11 (An easier-to-use version of convergence) *Given a sequence $\{(M_j, d_j(\cdot,\cdot), p_j)\}$ of pointed metric spaces which converges to the pointed metric space $(M, d(\cdot,\cdot), p)$ in the sense of Definition 8.9, we can find mappings $\phi_j : M \to M_j$ and $\psi_j : M_j \to M$ for each choice of j which satisfy the following properties. (Note that these mappings are not required to be continuous.) The mappings respect the basepoints, so that $\phi_j(p) = p_j$ and $\psi_j(p_j) = p$ for all j. The ϕ_j's and ψ_j's are approximate inverses of each other, in the sense that*

$$\lim_{j\to\infty} \sup\{d(\psi_j(\phi_j(x)), x) : x \in B_M(p,R)\} = 0, \tag{8.8}$$

$$\lim_{j\to\infty} \sup\{d_j(\phi_j(\psi_j(u)), u) : u \in B_{M_j}(p_j,R)\} = 0 \tag{8.9}$$

for all $R > 0$. These mappings are also approximate isometries, in the sense that

$$\lim_{j\to\infty} \sup\{|d_j(\phi_j(x), \phi_j(y)) - d(x,y)| : x,y \in B_M(p,R)\} = 0, \tag{8.10}$$

$$\lim_{j\to\infty} \sup\{|d_j(\psi_j(u), \psi_j(v)) - d_j(u,v)| : u,v \in B_{M_j}(p_j,R)\} = 0 \tag{8.11}$$

for every $R > 0$.

This is largely a matter of unwinding definitions. Let $\{(M_j, d_j(\cdot,\cdot))\}$ and $(M, d(\cdot,\cdot))$ be as above. The convergence assumption means that we have bilipschitz embeddings $f_j : (M_j, d_j(\cdot,\cdot)^\alpha) \to (\mathbf{R}^n, |x-y|)$ and $f : (M, d(\cdot,\cdot)^\alpha) \to (\mathbf{R}^n, |x-y|)$ such that $f_j(M_j)$ converges to $f(M)$ as subsets of $\mathbf{R}^n$. We also have the normalization $f_j(p_j) = f(p) = 0$. We choose $\phi_j : M \to M_j$ and $\psi_j : M_j \to M$ in such a way that

$$|f_j(\phi_j(x)) - f(x)| = \mathrm{dist}_{\mathbf{R}^n}(f(x), f_j(M_j)) \tag{8.12}$$

holds for all $x \in M$. That is, we choose $\phi_j(x) \in M_j$ so that it lies as close to x as possible, but this statement does not make sense until we use the embeddings into $\mathbf{R}^n$. Similarly we choose $\psi_j(z) \in M$ for $z \in M_j$ so that $\psi_j(z)$ lies as close to z as possible, in the sense that

$$|f(\psi_j(z)) - f_j(z)| = \mathrm{dist}_{\mathbf{R}^n}(f_j(z), f(M))). \tag{8.13}$$

The values of ϕ_j and ψ_j are not uniquely determined by these conditions, but we do not mind. We do ask that ϕ_j and ψ_j respect the basepoints, which is compatible with (8.12) and (8.13). These mappings are also typically discontinuous, but we do not mind that either.

These choices together with the bilipschitz conditions on the embeddings f_j, f imply that

$$d_j(\phi_j(x), p_j) \leq C\, d(x,p) \qquad \text{when } x \in M, \tag{8.14}$$

and

$$d(\psi_j(u), p) \leq C\, d_j(u, p_j) \qquad \text{when } u \in M_j, \tag{8.15}$$

where C does not depend on j or the point in question. Our convergence conditions imply that

$$\lim_{j\to\infty} \sup\{|f_j(\phi_j(x)) - f(x)| : x \in B_M(p,R)\} = 0 \tag{8.16}$$

$$\lim_{j\to\infty} \sup\{|f(\psi_j(u)) - f_j(u)| : u \in B_{M_j}(p_j,R)\} = 0 \tag{8.17}$$

for all $R > 0$. It is not hard to derive (8.8), (8.9) from (8.14)–(8.17). (Take $u = \phi_j(x)$ in (8.17) to conclude that $|f(\psi_j\phi_j(x)) - f(x)|$ is small, then use the bilipschitz condition on f to get that $d(\psi_j\phi_j(x), x)$ is small, etc.)

To prove the almost-isometry properties (8.10), (8.11) we use the convergence of the metrics provided by Definition 8.9. That is, the functions $d_j(f_j^{-1}(x), f_j^{-1}(y))$ on $f_j(M_j) \times f_j(M_j)$ converge to $d(f^{-1}(x), f^{-1}(y))$ on $f(M) \times f(M)$ in the sense of Definition 8.4. Lemma 8.7 implies a uniform version of this convergence, as in Remark 8.10. Once we have Lemma 8.7 it is easy to verify (8.10), (8.11) using (8.16), (8.17).

This completes the proof of Lemma 8.11.

Lemma 8.12 (Uniqueness of limits) *Let $\{(M_j, d_j(\cdot,\cdot), p_j)\}$ be a sequence of pointed metric spaces which converges in the sense of Definition 8.9 to each of the pointed metric spaces $(M, d(\cdot,\cdot), p)$ and $(N, D(\cdot,\cdot), q)$ (with possibly different embeddings and so forth). Then $(M, d(\cdot,\cdot), p)$ and $(N, D(\cdot,\cdot), q)$ are isometrically equivalent, with an isometry that respects the basepoints.*

This follows easily from Lemma 8.11. From there we have sequences of almost-isometries between M and the M_j's for large j, and between N and the M_j's for large j. We can compose them to get almost-isometries between M and N which respect the basepoints. To get an actual isometry we pass to a limit. More precisely, an Arzela–Ascoli-type argument implies that there is a convergent subsequence (by first finding a subsequence which converges on a countable dense subset), and it is not hard to check that the limiting map provides an isometry from M onto N which respects the basepoints. We omit the details.

Lemma 8.13 (Existence of limits) *Let $\{(M_j, d_j(\cdot,\cdot), p_j)\}$ be a sequence of pointed metric spaces which are complete and doubling, and with a uniform bound on their doubling constants. Fix a family of L-bilipschitz embeddings*

$$f_j : (M_j, d_j(\cdot,\cdot)^\alpha) \to (\mathbf{R}^n, |x-y|), \tag{8.18}$$

where $\alpha \in (0,1]$, and where L, α, and n do not depend on j, and assume also that $f_j(p_j) = 0$ for all j. (The existence of such a family of mappings is provided by Assouad's embedding theorem, at least when $\alpha < 1$ and n and L are large enough.) Then by passing to a subsequence we get convergence to a pointed metric space $(M, d(\cdot,\cdot), p)$ in the same manner as in Definition 8.9 (with these choices of α, n, and the f_j's).

This is pretty straightforward. By Lemma 8.2 we may assume that $f_j(M_j)$ converges to a nonempty closed set $F \subseteq \mathbf{R}^n$ in the sense of Definition 8.1. We also have convergence of $f_j(M_j) \times f_j(M_j)$ to $F \times F$, as in Remark 8.3. The functions $d_j(f_j^{-1}(x), f_j^{-1}(y))^\alpha$ on $f_j(M_j) \times f_j(M_j)$ are automatically equicontinuous and uniformly bounded on bounded subsets in the present circumstances, as in Remark 8.10. Thus we may apply Lemma 8.6 to conclude the existence of a limit after passing to a subsequence. We define $d(x,y)$ on $F \times F$ so that $d(x,y)^\alpha$ is this limit. It is not hard to prove that $d(x,y)$ is actually a metric, using the corresponding property of $d_j(f_j^{-1}(x), f_j^{-1}(y))$. Also, $d(x,y)^\alpha$ is equivalent in size to $|x-y|^\alpha$, and in fact

$$L^{-1}\,|x-y| \le d(x,y)^\alpha \le L\,|x-y| \tag{8.19}$$

for all $x, y \in F$, because of convergence and the corresponding property for $d_j(f_j^{-1}(x), f_j^{-1}(y))^\alpha$. Altogether we take $M = F$ with this metric $d(x,y)$, $p = 0$, and we take the embedding $f : M \to \mathbf{R}^n$ taken to be the identity. This proves the lemma.

Remark 8.14 (Compatibility with convergence of subsets) We now have two notions of convergence for subsets of Euclidean spaces, namely Definitions 8.1 and 8.9. They are compatible in the following way. If a sequence converges in both senses, then the limits are isometric by the uniqueness result Lemma 8.12. Convergence as subsets of Euclidean spaces implies convergence as metric spaces (modulo the minor issue of basepoints), but the reverse need not be true. However, we know that we can always get convergent subsequences, because of Lemma 8.2, and then Lemma 8.12 determines the geometry of all possible subsequential limits up to isometry. Thus the two notions are practically equivalent in terms of the limits that they give if one is willing to pass to subsequences, and modulo the small issue of basepoints.

Remark 8.15 (Dependence on the free parameters) In Definition 8.9 we had some free parameters, namely α, the dimension n of the receiving Euclidean space, and the embeddings. We would like to say that these choices do not matter, that any general limit could also be obtained with particular choices of embeddings, and so forth. This is not true, because we could have convergence for one sequence of embeddings but not for another. However, Lemma 8.13 tells us that we can always get existence by passing to a subsequence. This means that whatever limit we can obtain in general can also be obtained from any particular sequence of embeddings, if we are willing to pass to a subsequence. This assertion uses Lemma 8.12 as well.

8.5 Convergence of mappings between spaces

We shall need to extend our notion of convergence of mappings to maps between metric spaces that are moving themselves. We shall use "mapping packages" to encorporate the domain, range, and the mapping itself.

Definition 8.16 (Mapping packages) *A* mapping package *consists of a pair of pointed metric spaces* $(M, d(x,y), p)$, $(N, \rho(u,v), q)$ *and a mapping* $g : M \to N$ *between them which respects the basepoints, i.e.,* $g(p) = q$.

Definition 8.17 (Equicontinuity and uniform boundedness) *Given a sequence* $\{(M_j, d_j(x,y), p_j), (N_j, \rho_j(u,v), q_j), h_j\}$ *of mapping packages, we say that the* h_j*'s are equicontinuous on bounded subsets if for each* $R > 0$ *and* $\epsilon > 0$ *there exists* $\delta > 0$ *so that*

$$\rho_j(h_j(x), h_j(y)) < \epsilon \qquad \text{when } x, y \in B_{M_j}(p_j, R) \text{ and } d_j(x,y) < \delta. \tag{8.20}$$

We say that the h_j*'s are* uniformly bounded on bounded sets *if*

$$\sup_j \sup_{x \in B_{M_j}(p_j, R)} \rho_j(h_j(x), q_j) < \infty \tag{8.21}$$

for all $R > 0$.

Note that these properties are automatic when the h_j's are Lipschitz with a bounded constant, since the h_j's respect the basepoints. In practice we shall typically have such a uniform bound.

Definition 8.18 (Convergence of mapping packages) *Given a sequence of mapping packages* $\{(M_j, d_j(x,y), p_j), (N_j, \rho_j(u,v), q_j), h_j\}$, *we say that a mapping package* $\{(M, d(x,y), p), (N, \rho(u,v), q), h\}$ *is the limit of this sequence if the following conditions are satisfied. All the metric spaces should be complete and doubling. In addition we ask for embeddings* f_j, f *of the* $(M_j, d_j(x,y), p_j)$*'s and* $(M, d(x,y), p)$ *into some* $\mathbf{R}^n$, *and for embeddings* g_j, g *of the* $(N_j, \rho_j(u,v), q_j)$*'s and* $(N, \rho(u,v), q)$ *into some* $\mathbf{R}^l$, *with entirely the same properties as in Definition 8.9. We also ask that the mappings* $g_j \circ h_j \circ f_j^{-1}$ *converge to* $g \circ h \circ f^{-1}$ *in the sense of Definition 8.4.*

Keep in mind that the embeddings that we use contain implicitly a choice of exponent α. We are allowed to use a different choice for the N's than for the M's (but of course we use the same choice for all the M's and all the N's).

For this notion of convergence we have properties like the ones before.

Lemma 8.19 (An easier-to-use version of convergence) *Let*

$$\{(M_j, d_j(x,y), p_j), (N_j, \rho_j(u,v), q_j), h_j\}$$

be a sequence of mapping packages which converges to the mapping package $\{(M, d(x,y), p), (N, \rho(u,v), q), h\}$ *as in the preceding definition. Then we can find* *sequences of mappings $\phi_j : M \to M_j$ and $\psi_j : M_j \to M$ which satisfy exactly the same conditions as in Lemma 8.11, and sequences of mappings $\sigma_j : N \to N_j$ and $\tau_j : N_j \to N$ which satisfy similar conditions as in Lemma 8.11, in such a way that we also have*

$$\lim_{j\to\infty} \tau_j(h_j(\psi_j(x))) = h(x) \tag{8.22}$$

for all $x \in M$. If the mappings h_j are also equicontinuous and uniformly bounded on bounded sets, then $\tau_j \circ h_j \circ \psi_j : M \to N$ converges to h uniformly on bounded subsets of M.

This is not hard to check, and we only sketch the argument. The main point is that the mappings ϕ_j's and ψ_j's given by Lemma 8.11 were chosen according to the "nearest point" rules (8.12), (8.13) (and similarly for the σ_j's and τ_j's). This permits the pointwise convergence property (8.22) to be deduced from its counterpart in Definition 8.4 (which is part of the definition of convergence of mapping packages). The statement about uniform convergence follows in the same way from Lemma 8.7.

Lemma 8.20 (Lipschitz and bilipschitz conditions) *Let*

$$\{(M_j, d_j(x,y), p_j), (N_j, \rho_j(u,v), q_j), h_j\}$$

be a sequence of mapping packages which converges to the mapping package $\{(M, d(x,y), p), (N, \rho(u,v), q), h\}$. If all the h_j's are L-Lipschitz for some L, then h is too. If all the h_j's are L-bilipschitz, then the same is true of h.

This is easy to check using the preceding lemma. The main point is to go back to (8.10), (8.11).

Lemma 8.21 (Uniqueness) *If $\{(M_j, d_j(x,y), p_j), (N_j, \rho_j(u,v), q_j), h_j\}$ is a sequence of mapping packages which converges to both of the mapping packages*

$$\{(M, d(x,y), p), (N, \rho(u,v), q), h\}, \qquad \{(M', d'(x,y), p'), (N', \rho'(u,v), q'), h'\},$$

then there are isometries between M and M', and between N and N', which respect the basepoints, and which intertwine h and h' in the obvious way.

This is an easy consequence of Lemma 8.19, just as for Lemma 8.12.

Lemma 8.22 (Existence) *Let*

$$\{(M_j, d_j(x,y), p_j), (N_j, \rho_j(u,v), q_j), h_j\}$$

be a sequence of mapping packages, where all the metric spaces are complete and doubling (with uniformly bounded doubling constants), and where the mappings h_j are equicontinuous and uniformly bounded on bounded sets. Fix families of L-bilipschitz embeddings

$$f_j : (M_j, d_j(\cdot,\cdot)^\alpha) \to (\mathbf{R}^n, |x-y|), \qquad g_j : (N_j, \rho_j(\cdot,\cdot)^\beta) \to (\mathbf{R}^l, |x-y|),$$

where $\alpha, \beta \in (0,1]$, and where L, α, β, n, and l do not depend on j, and assume also that $f_j(p_j) = 0$, $g_j(q_j) = 0$ for all j. (The existence of such a family of *mappings is guaranteed by Assouad's embedding theorem, at least when $\alpha, \beta < 1$.) Then by passing to a subsequence we get convergence to a mapping package as in Definition 8.18.*

To prove this one simply applies Lemma 8.13 to get convergence of the relevant pointed metric spaces, and then Lemma 8.6 to get the convergence of the mappings. There are small details to check, but they are easy.

One can make the same kind of observations about the dependence of the limits of mapping packages on the embeddings, etc. used as for convergence of spaces, as in Remark 8.15. In other words, there is practically no dependence, if we are willing always to pass to subsequence, and if we work up to isometric equivalence.

8.6 Convergence of measures

As usual we begin with convergence on Euclidean spaces before proceeding to the general case.

Definition 8.23 (Convergence of measures on Euclidean spaces) *Let $\{\mu_j\}$ be a sequence of nonnegative Borel measures on some $\mathbf{R}^n$ which are finite on compact sets, and let μ be another measure on $\mathbf{R}^n$ with the same properties. We say that $\{\mu_j\}$* converges *to μ if*

$$\lim_{j\to\infty} \int_{\mathbf{R}^n} \phi \, d\mu_j = \int_{\mathbf{R}^n} \phi \, d\mu \tag{8.23}$$

for every continuous function ϕ on $\mathbf{R}^n$ with compact support.

Strictly speaking we ought to say something like "weak convergence" to be compatible with standard terminology, but for simplicity we shall not bother.

Note that the limit of a sequence of measures is unique when it exists. For this it is helpful to notice that these measures must be Borel regular, as in Theorem 2.18 of [R].

Lemma 8.24 (Existence of limits for measures) *Let $\{\mu_j\}$ be a sequence of nonnegative Borel measures on some $\mathbf{R}^n$, and suppose that*

$$\sup_j \mu_j(B(0,R)) < \infty \tag{8.24}$$

for all $R > 0$. Then there is a subsequence of $\{\mu_j\}$ which converges.

This is well known. One uses a Cantor diagonalization argument, reducing first to showing that (8.23) exists for a countable collection of ϕ's. The limiting measure is realized first as a nonnegative linear functional on the continuous functions with compact support, and then as a measure using the Riesz representation theorem.

Note that (8.24) is necessary for the existence of a limit.

We can extend this notion of convergence of measures to converging sequences of metric spaces in the obvious way. Here is the definition.

Definition 8.25 (Convergence of measures on general spaces) *Suppose that $\{(M_j, d_j(\cdot,\cdot), p_j)\}$ is a sequence of pointed metric spaces which converges to $(M, d(\cdot,\cdot), p)$ in the sense of Definition 8.9, with n, $\{f_j\}$, f, α as in Definition 8.9. Assume also that we are given a nonnegative Borel measure μ_j on each M_j, and another one μ on M, all of which are finite on compact sets. We say that the μ_j's converge to μ (with respect to these choices of embeddings) if their push-forwards to $\mathbf{R}^n$ using the f_j's and f converge in the sense of Definition 8.23.*

Lemma 8.26 (Existence of limits for measures) *Let $\{(M_j, d_j(\cdot,\cdot), p_j)\}$ be a sequence of pointed metric spaces which converges to $(M, d(\cdot,\cdot), p)$ in the sense of Definition 8.9, and suppose that for each j we have a nonnegative Borel measure μ_j on M_j such that*

$$\sup_j \mu_j(B_{M_j}(0, R)) < \infty \tag{8.25}$$

for all $R > 0$. Then we can get convergence of the measures after passing to a subsequence.

This follows easily from Lemma 8.24.

It will be convenient for us to have the notion of "Ahlfors regularity" for measures, and to know about their limits.

Definition 8.27 (Ahlfors regular measures) *Let M be a complete metric space. We say that a nonnegative Borel measure μ is* Ahlfors regular of dimension *d if there exists a constant $C > 0$ such that*

$$C^{-1}\, r^d \le \mu(B(x,r)) \le C\, r^d \tag{8.26}$$

whenever $0 < r \le \operatorname{diam} M$ and x lies in the support of μ. (In the present generality we can define "the support of μ" to be the set of points $z \in M$ such that $\mu(B(z,t)) > 0$ for all $t > 0$.)

Notice that the support of an Ahlfors regular measure is an Ahlfors regular set, because of Lemma 1.2.

Lemma 8.28 (Limits of Ahlfors regular measures) *Suppose that we have a sequence of pointed metric spaces converging to another pointed metric space (in the sense of Definition 8.9), and also a sequence of nonnegative Borel measures converging to a nonnegative Borel measure on the limiting space, as in Definition 8.25. If the measures in the sequence are Ahlfors regular with the same dimension d and a bounded constant C, then the limiting measure is either the zero measure or Ahlfors regular with dimension d and constant C. If the measures in the sequence have the whole metric space as their support, then the limiting measure has the same property.*

This is easy to check. The main point is that the definitions are set up in such a way that this fact almost reduces to its counterpart on $\mathbf{R}^n$. To get the right

Ahlfors regularity constant one has to be more careful and use the convergence of the distance functions. Convergence to the zero measure occurs precisely when the distance between the supports of the measures and the basepoints tends to infinity. The last part is easy to check because the Ahlfors regularity condition prevents the measures from becoming too thin anywhere. Indeed if one has a sequence of Ahlfors regular measures on $\mathbf{R}^n$ with bounded constants converging to a nonzero measure, then the supports of the measures converge to the support of the limit as subsets of $\mathbf{R}^n$. This is not hard to verify.

Lemma 8.29 (Limits of Ahlfors regular spaces) *Let $(M_j, d_j(x,y), p_j)$ be a sequence of Ahlfors regular pointed metric spaces of dimension d, with bounded regularity constants, and suppose that $\{(M_j, d_j(x,), p_j)\}$ converges to the pointed metric space $(M, d(x,y), p)$ in the sense of Definition 8.9. Then $(M, d(x,y))$ is Ahlfors regular, with the same dimension, and with a constant which is controlled by the regularity constants for the $(M_j, d_j(x,))$'s.*

This follows from Lemma 8.28. More precisely, to say that the M_j's are Ahlfors regular means that the restriction of d-dimensional Hausdorff measure to each of them is a regular measure, supported on the whole space M_j, and with bounded regularity constant. Using Lemma 8.26 we can pass to a subsequence to reduce to the case where these Hausdorff measures converge to a measure on the limiting space. The limiting measure is Ahlfors regular of dimension d and is supported on the whole limiting space, by Lemma 8.28. From Lemma 1.2 we conclude that the limiting space is Ahlfors regular, as desired.

8.7 Limits of subsets and their measures

Given a sequence of pointed metric spaces which converges to some other metric space, it will be convenient for us to be able to speak of convergence of subsets of the spaces in the sequence. We use the following definition.

Definition 8.30 (Convergence of subsets of general spaces) *Assume that $\{(M_j, d_j(\cdot,\cdot), p_j)\}$ is a sequence of pointed metric spaces which converges to the pointed metric space $\{(M, d(\cdot,\cdot), p)\}$ in the sense of Definition 8.9, and that embeddings f_j, f into some $\mathbf{R}^n$ have been specified, etc., as in Definition 8.9. Assume also that we are given a closed subset F_j of M_j for each j, and also a closed subset F of M, all nonempty. We say that the F_j's converge to F if their images $f_j(F_j)$ in $\mathbf{R}^n$ converge to $f(F)$ in the sense of Definition 8.1.*

Thus the convergence depends *a priori* on the embeddings used, but different choices of embeddings yield equivalent answers, in the sense that limits obtained from different choices are isometrically equivalent, as in Lemma 8.12. More precisely, if we have another candidate limiting space M' with another limiting subset F', then there is an isometry from M onto M' which maps F onto F'.

Lemma 8.31 (Existence of limits of subsets of general spaces) *Suppose that $\{(M_j, d_j(\cdot,\cdot), p_j)\}$ is a sequence of pointed metric spaces which converges to* *the pointed metric space $\{(M, d(\cdot,\cdot), p)\}$ in the sense of Definition 8.9, with some given choice of embeddings into a Euclidean space, etc. Let $\{F_j\}$ be a sequence of nonempty closed subsets of the M_j's, and assume that*

$$\sup_j \operatorname{dist}(F_j, p_j) < \infty. \tag{8.27}$$

Then we can pass to a subsequence to get convergence to a nonempty closed subset F of M.

This is an easy consequence of the definitions and Lemma 8.2.

In practice we shall sometimes be dealing with subsets which arise as images of other sets under a convergent sequence of mappings, and the next lemma provides us with the relevant convergence result.

Lemma 8.32 (Convergence of images) *Let*

$$\{(M_j, d_j(x,y), p_j), (N_j, \rho_j(u,v), q_j), h_j\} \tag{8.28}$$

be a sequence of mapping packages which converges to the mapping package $\{(M, d(x,y), p), (N, \rho(u,v), q), h\}$. Assume that the h_j's are uniformly bounded and equicontinuous on bounded sets. (See Definitions 8.18 and 8.17.) Let $\{K_j\}$ be a sequence of compact subsets of the M_j's which converges to $K \subseteq M$ in the sense of Definition 8.30, and assume that

$$\sup_j \{\operatorname{diam}_{M_j} K_j + \operatorname{dist}_{M_j}(K_j, p_j)\} < \infty \tag{8.29}$$

(to avoid pathologies like a piece of K_j leaking off to infinity). Then $\{h_j(K_j)\}$ converges as subsets of N_j to $h(K) \subseteq N$.

This is easy to check. The main point is to unwind the definitions to reduce to the case where all the M's and N's are subsets of some $\mathbf{R}^n$, and then to use the uniform convergence of mappings provided by Lemma 8.7.

We shall need to know what happens to the measures of subsets in the limit. It is easy to make examples of sequences of finite sets which fill up a ball in the limit, so that measures can get bigger in the limit. It is not so easy for them to get smaller, however.

Lemma 8.33 (Limits of measures of compact sets) *Let $\{(M_j, d_j(\cdot,\cdot), p_j)\}$ be a sequence of pointed metric spaces which converges to the pointed metric space $\{(M, d(\cdot,\cdot), p)\}$ in the sense of Definition 8.9. Let $\{K_j\}$ be a sequence of nonempty compact subsets of the M_j's which converges to a nonempty compact set K in M in the sense of Definition 8.30, and assume also that*

$$\sup_j \{\operatorname{diam}_{M_j} K_j + \operatorname{dist}_{M_j}(K_j, p_j)\} < \infty. \tag{8.30}$$

Let $\{\mu_j\}$ be a sequence of nonnegative Borel measures on the M_j's which converges to a Borel measure μ on M in the sense of Definition 8.25. Then

$$\mu(K) \geq \limsup_{j \to \infty} \mu_j(K_j). \tag{8.31}$$

Let us prove the lemma. We may as well assume that all of our metric spaces M_j lie in $\mathbf{R}^n$ already, since the effect of our definitions is to reduce everything to that case. Thus we can simply think that our measures μ_j, μ already live on $\mathbf{R}^n$, and that the K_j's and K are compact subsets of $\mathbf{R}^n$. All of our measures μ_j and μ are supposed to be finite on compact sets, as in Definition 8.25. This implies that they are Borel regular, as in Theorem 2.18 in [R]. As a consequence for each $\epsilon > 0$ we can find an open set U of $\mathbf{R}^n$ containing K such that

$$\mu(U) < \mu(K) + \epsilon. \tag{8.32}$$

Let ϕ be a continuous function with compact support on $\mathbf{R}^n$ such that $0 \leq \phi \leq 1$ everywhere, $\phi \equiv 1$ on a neighborhood of K, $\phi \equiv 0$ on the complement of U. Then $\phi \equiv 1$ on K_j for all sufficiently large j (since the K_j's converge to K, and there is nothing leaking off to infinity), from which we conclude that

$$\mu(K) + \epsilon > \int_{\mathbf{R}^n} \phi \, d\mu = \lim_{j \to \infty} \int_{\mathbf{R}^n} \phi \, d\mu_j \geq \limsup_{j \to \infty} \mu_j(K_j). \tag{8.33}$$

The lemma follows, since $\epsilon > 0$ is arbitrary.

Next let us consider what happens to Hausdorff measures in the limit. We cannot say directly that the Hausdorff measures always increase in the limit, we need to make an assumption.

Definition 8.34 (Ahlfors subregularity) *A metric space A is* Ahlfors subregular *of dimension d if*

$$H^d(A \cap B(x,r)) \leq C\, r^d \tag{8.34}$$

for some constant C and all $x \in A$, $r > 0$.

Subsets of Ahlfors regular spaces are automatically Ahlfors subregular.

Lemma 8.35 *Let $\{(M_j, d_j(x,y), p_j)\}$ be a sequence of pointed metric spaces which converges to $(M, d(x,y), p)$ in the sense of Definition 8.9. Suppose that $\{A_j\}$ is a sequence of compact subsets of the M_j's which converges to the compact set A in M, and that*

$$\sup_j \{\operatorname{diam}_{M_j} A_j + \operatorname{dist}_{M_j}(A_j, p_j)\} < \infty. \tag{8.35}$$

Suppose also that the A_j's are Ahlfors subregular with the same dimension d and a uniformly bounded constant. Then

$$H^d(A) \geq C^{-1} \limsup_{j \to \infty} H^d(A_j), \tag{8.36}$$

where C depends only on d and the bounds on the subregularity constants for the A_j's.

In our applications the M_j's will typically be Ahlfors regular themselves, with uniformly bounded constants, and one can obtain the conclusions of this lemma from Lemma 8.33. Still, it is pleasant to give a direct argument.

Let us prove Lemma 8.35. We may as well assume that $H^d(A) < \infty$. By definition of Hausdorff measure we have that for each $\epsilon > 0$ there is a covering of A by a sequence $\{E_i\}$ of subsets of A with $\sum_i (\operatorname{diam}_M E_i)^d < H^d(A) + \epsilon$. We may assume that the E_i's are actually open subsets of A, because we can always increase them slightly without disturbing the content of the estimate. The compactness of A implies then that we may reduce to a finite subcovering. Using this finite covering it is not hard to show that for each sufficiently large j we can find a finite collection of subsets $\{E_{i,j}\}_i$ of A_j such that A_j is covered by $\{E_{i,j}\}_i$ and

$$\sum_i (\operatorname{diam}_{M_j} E_{i,j})^d < H^d(A) + 2\,\epsilon. \tag{8.37}$$

This can be accomplished by adapting (or interpreting) Lemma 8.11 to the present context of convergence of subsets. Ahlfors subregularity permits us to conclude that

$$H^d(A_j) \le \sum_i H^d(E_{i,j}) \le C \sum_i (\operatorname{diam}_{M_j} E_{i,j})^d. \tag{8.38}$$

From here Lemma 8.35 follows easily.

Let us use this to give another proof of Lemma 8.29. Let $(M_j, d_j(x,), p_j)$, $(M, d(x,y), p)$ be as in the lemma. The definition of convergence ensures that $(M, d(x,y))$ is complete. We need only check the upper and lower bounds for the Hausdorff measure of balls. We begin with the lower bound.

Fix $x \in M$ and $0 < r \le \operatorname{diam} M$. Using the definition of convergence (as in Lemma 8.11 and its proof) it is not difficult to find points $x_j \in M_j$ such that, after passing to a subsequence, we have that $\overline{B}_{M_j}(x_j, r)$ converges to a subset of $\overline{B}_M(x,r)$. Here we use convergence in the sense of Definition 8.30, and we can get the existence of the subsequential limit from Lemma 8.31. Note that we do have to be slightly careful here, we may not get the whole ball in the limit, something could leak in from the outside. We could avoid this by increasing the radii slightly, but we are happy enough to take all the radii to be the same and then miss something in the limit. We obtain that

$$H^d(\overline{B}_M(x,r)) \ge C^{-1} \liminf_{j \to \infty} H^d(\overline{B}_{M_j}(x_j, r)) \tag{8.39}$$

for a suitable constant C, because of Lemma 8.35. Hence

$$H^d(\overline{B}_M(x,r)) \ge C^{-1}\, r^d \tag{8.40}$$

for another constant C, because of the Ahlfors regularity of the M_j's. (We are implicitly using also the simple fact that $\operatorname{diam} M = \lim_{j\to\infty} \operatorname{diam} M_j$.)

This gives us the lower bound for regularity, and for the upper bound we use Lemma 5.1. That is, Lemma 5.1 applies to each of the M_j's, with a uniform bound, and it is easy to see that the property in the conclusion of Lemma 5.1 behaves well under limits of metric spaces, so that it holds for M too. The covering property in the conclusion of Lemma 5.1 implies easily the upper bound for Ahlfors regularity, straight from the definition of Hausdorff measure. Thus we get

$$H^d(B_M(x,r)) \leq C\,r^d \tag{8.41}$$

for some constant C and all $x \in M$, $0 < r \leq \operatorname{diam} M$, as desired.

8.8 Smooth sets

In general the measures of sets can jump up in the limit. In this section we give a condition under which this does not happen. Namely one should require a kind of "smoothness" of the subsets.

Definition 8.36 (Smooth sets) *Let M be a metric space, and let $A, d,$ and t be positive numbers. A subset K of M is said to be A-*smooth of dimension d and size t *if for each $0 < r \leq t$ the sets*

$$\{y \in M\backslash K : \operatorname{dist}(y,K) < r\}, \qquad \{y \in K : \operatorname{dist}(y, M\backslash K) < r\} \tag{8.42}$$

can be covered by $A\,(\frac{t}{r})^{d-A^{-1}}$ balls of radius r.

This is a way of saying that K "has small boundary" in M. It implies that

$$H^d(\{y \in M\backslash K : \operatorname{dist}(y,K) < r\}) \leq 2^d\,A\,(\frac{r}{t})^{A^{-1}}\,t^d \tag{8.43}$$

and

$$H^d(\{y \in K : \operatorname{dist}(y, M\backslash K) < r\}) \leq 2^d\,A\,(\frac{r}{t})^{A^{-1}}\,t^d \tag{8.44}$$

when $0 < r \leq t$, by definition of Hausdorff measure. One should think of t as representing the "size" of the boundary of K (something like its diameter).

In practice M will be Ahlfors regular of dimension d and we shall use the same dimension d for the smoothness condition. If K itself has diameter t, for instance, then we can automatically cover the sets in (8.42) with $O((t/r)^d)$ balls of radius r. The point is to have coverings with a smaller number of balls. One can show that bounds on measures (8.43), (8.44) are equivalent to the covering condition in the Ahlfors regular case (with a bounded distortion of the constants.) (This is similar to Lemma 5.1.)

If M is $\mathbf{R}^d$, then balls and cubes are smooth sets. In this case we can cover the analogues of the sets (8.42) with $O((t/r)^{d-1})$ balls of radius r. Any bounded smooth or (finitely) polyhedral domain in $\mathbf{R}^d$ enjoys the same property. In general our definition permits domains which are not smooth in the classical sense, which are even fractal, but for simplicity we stick to this terminology.

One can obviously tinker with the definition of a smooth set. We have chosen this one so that it comes with scale-invariant estimates, and so that it is as strong as possible subject to the constraints of what we can get in practice. It would be reasonable to consider weakening the condition that we win a positive power of t/r in the number of balls in the covering condition, any function which goes to infinity with t/r would work just as well. This would be sufficient for the application to the limits of measures, and for that we only need to control the part of the complement of K near K and not the other part.

It is not immediately obvious that there are plenty of smooth sets under general conditions. There is no reason *a priori* why balls should be smooth in a general space, for instance. Fortunately we can use cubes (Section 5.5) on Ahlfors regular spaces.

Lemma 8.37 (Cubes are smooth sets) *Let M be a regular metric space of dimension d. Then every cube in M (in the sense of Section 5.5) is A-smooth with dimension d and size equal to its diameter, with a constant A that depends only on the constants in (5.6)–(5.9).*

Recall from Section 5.5 that one can always find cubes on an Ahlfors regular space whose constants are controlled by the regularity constants.

The proof of the lemma is an easy exercise. The main point behind the bounds on the coverings of (8.42) is (5.9). That provides bounds on the mass like (8.43), (8.44), which is not quite what we want, but one can convert them into bounds on coverings without much trouble. (If the set in question is a cube of size 2^j, then it is best to think of (8.42) with $r = 2^l$, $l \leq j$, and to cover (8.42) by the cubes of size 2^l which touch them. These cubes are pairwise disjoint, and so their number is controlled by the total mass, etc.) This proves the lemma.

Lemma 8.38 (Unions, intersections, and complements) *Complements of A-smooth sets of dimension d and size t are also smooth with the same parameters. Finite unions and intersections are also smooth, but perhaps with an increase in the value of A.*

This is easy to check from the definitions.

Lemma 8.39 (Limits of measures of smooth sets) *Let $\{(M_j, d_j(\cdot,\cdot), p_j)\}$ be a sequence of pointed metric spaces which converges to the pointed metric space $\{(M, d(\cdot,\cdot), p)\}$ in the sense of Definition 8.9. Let $\{K_j\}$ be a sequence of nonempty compact subsets of the M_j's which converges to a nonempty compact set K in M in the sense of Definition 8.30, and assume also that*

$$\sup_j \{\operatorname{diam}_{M_j} K_j + \operatorname{dist}_{M_j}(K_j, p_j)\} < \infty \tag{8.45}$$

(to avoid pathologies like a piece of K_j leaking off to infinity). Let $\{\mu_j\}$ be a sequence of nonnegative Borel measures on the M_j's which converges to a Borel measure μ on M in the sense of Definition 8.25. Assume also that there exist constants constants A, d, t such that the K_j's are all A-smooth with dimension d

and size t, and such that the measures μ_j are all Ahlfors regular with dimension d and constant A. Then

$$\mu(K) = \lim_{j\to\infty} \mu_j(K_j). \tag{8.46}$$

For this we do not need the full strength of Ahlfors regularity, only the upper bound in (8.26).

To prove the lemma we use approximately the same argument as for Lemma 8.33. We first observe that the definitions permit us to reduce to the case where the M_j's and M are contained in some $\mathbf{R}^n$. More precisely, the assumptions of the smoothness of the sets and the Ahlfors regularity of the measures are not disturbed by applying snowflake transforms to our spaces or using bilipschitz embeddings. The snowflake transforms will lead to a change in the dimensions, but we do not mind, the dimensions will all be changed in the same way. The constants will change, but in a bounded way.

Thus we assume that M and the M_j's all lie in $\mathbf{R}^n$.

Given $r > 0$, set $K(r) = \{x \in \mathbf{R}^n : \operatorname{dist}(x, K) \leq r\}$. Fix an r, small. Let ϕ be a continuous function on $\mathbf{R}^n$ with compact support such that $0 \leq \phi \leq 1$ everywhere, $\phi \equiv 1$ on $K(r)$ and $\phi \equiv 0$ off of $K(2r)$. Our assumption of convergence of measures implies that

$$\int_{\mathbf{R}^n} \phi \, d\mu = \lim_{j\to\infty} \int_{\mathbf{R}^n} \phi \, d\mu_j. \tag{8.47}$$

The main point now is that $\int_{\mathbf{R}^n} \phi \, d\mu$ is a good approximation for $\mu(K)$ and $\int_{\mathbf{R}^n} \phi \, d\mu_j$ is a good approximation for $\mu_j(K_j)$ when r is small and j is large. Indeed,

$$\left|\mu(K) - \int_{\mathbf{R}^n} \phi \, d\mu\right| \tag{8.48}$$

is small when r is small just by Borel regularity (which we can get as usual from Theorem 2.18 in [R]). Let us check that

$$\left|\mu_j(K_j) - \int_{\mathbf{R}^n} \phi \, d\mu_j\right| \leq C\, r^a \tag{8.49}$$

for some $a > 0$ and all sufficiently large j. How large j has to be depends on r, but the constants C, a do not depend on either j or r. To prove this estimate we observe first that

$$\mu_j(K_j) \leq \int_{\mathbf{R}^n} \phi \, d\mu_j \tag{8.50}$$

for all sufficiently large j, because $K_j \subseteq K(r)$ when j is large enough (since K_j converges to K). Thus it suffices to show that

$$\mu_j(K(2r) \backslash K_j) \leq C\, r^a \tag{8.51}$$

when j is large enough. Of course we need only estimate $\mu_j(K(2r) \cap M_j \backslash K_j)$, since μ_j is supported on M_j. The point now is that

$$K(2r) \cap M_j \backslash K_j \subseteq \{x \in M_j \backslash K_j : \operatorname{dist}(x, K_j) \leq 3r\} \tag{8.52}$$

when j is large enough. Indeed, every element of $K(2r)$ lies within $2r$ of K, and hence within $3r$ of K_j when j is large enough, by convergence. Thus we get the desired inclusion. The inequality (8.51) follows from the smoothness of K_j and the Ahlfors regularity of μ_j, with a constant that does not depend on j or r. This implies that (8.49) holds for each r and for sufficiently large j, with constants C, a that are independent of r and j. Combining this with (8.47) and the smallness of (8.48) for small r we get (8.46), as desired.

This completes the proof of the lemma.

9

WEAK TANGENTS

Now we want to apply the notions of convergence from Chapter 8 to BPI spaces. This will permit us to make some compactness arguments.

9.1 The definition

Let $(M, d(x,y))$ be a metric space. For each $p \in M$ and $0 < t \leq \operatorname{diam} M$ we use $M_{p,t}$ to mean the pointed metric space $(M, t^{-1}\, d(x,y), p)$. Notice that $M_{p,t}$ always has diameter ≥ 1.

Suppose that M is doubling and complete. We call a metric space a *weak tangent* of M if it arises as the limit of a sequence of $M_{p,t}$'s, where we use the notion of convergence given in Definition 8.9. The $M_{p,t}$'s themselves are all considered to be among the weak tangents (using constant sequences).

Lemma 8.13 implies that any sequence of (p,t)'s has a subsequence for which the $M_{p,t}$'s converge. This uses the simple observation that the $M_{p,t}$'s all have the same doubling constant.

We write $WT(M)$ for the collection of all weak tangents to M. Thus the elements of $WT(M)$ are pointed metric spaces which are doubling and complete.

Note that in defining the weak tangents of a metric space we may restrict the parameters α and n, as in Remark 8.15. We can start with any fixed bilipshitz embedding of $(M, d(x,y)^\alpha)$ into some $\mathbf{R}^n$ (using Assouad's embedding theorem if necessary), and then use translations and dilations of this fixed map for each $M_{p,t}$. We get all the weak tangents from limits of these embeddings as in Remark 8.15, because we do not mind passing to subsequences, and because we can use Lemma 8.13.

If M were simply a closed subset of some $\mathbf{R}^n$, then we might take for $M_{p,t}$ the set which is the image of M under the mapping $x \mapsto t^{-1}\,(x-p)$. This would give a subset of $\mathbf{R}^n$ isometric to $M_{p,t}$ as defined above, with the base point taken to be the origin. We might then wish to define the notion of weak tangent using convergence of subsets of Euclidean spaces, as in Definition 8.1, rather than Definition 8.9. If we did that we would get the same collection of weak tangents, up to isometric equivalence, as in Remark 8.14.

The reader may prefer to work only with subsets of Euclidean spaces, in order to avoid the complication of taking limits of abstract spaces. The resulting loss of generality is limited by Assouad's embedding theorem.

We might sometimes wish to write $WT_p(M)$ for the weak tangents to M that arise from limits of sequences of the form $\{M_{p,t_j}\}$ with $t_j \to 0$. That is, weak tangents obtained by blowing up at a particular point p.

Notice that if $(N, \rho(\cdot,\cdot)) \in WT(M)$, then

$$(N, \lambda\,\rho(\cdot,\cdot)) \in WT(M) \qquad \text{for all } \lambda \geq 1. \tag{9.1}$$

This is easy to check, by adjusting the t's in the approximating sequence of $M_{p,t}$'s. This is also true for all $\lambda > 0$ as long as either M is unbounded or M is bounded and N was obtained as the limit of $\{M_{p,t_j}\}$ with $t_j \to 0$. These conditions cover the cases of primary interest.

Similarly, if one changes the basepoint of an element of $WT(M)$, then one continues to have an element of $WT(M)$. This is not difficult to show by unwinding the definitions of convergence.

9.2 First facts

Definition 9.1 (Doubling condition for families of spaces) *Let $\mathcal{E}$ be a collection of metric spaces (possibly pointed). We say that $\mathcal{E}$ is* doubling *if each of the metric spaces in $\mathcal{E}$ is doubling, with a uniformly bounded constant.*

Lemma 9.2 *If M is a complete metric space and doubling, then $WT(M)$ is doubling as a family.*

This is not hard to check, from the definitions.

Definition 9.3 (Closed families of spaces) *A collection $\mathcal{E}$ of pointed complete metric spaces which is doubling is said to be* closed *if any limit of any sequence of elements of $\mathcal{E}$ (in the sense of Definition 8.9) also lies in $\mathcal{E}$. (Strictly speaking, we should probably identify metric spaces which are isometric, or ask only that the limit be isometrically equivalent to an element of $\mathcal{E}$.)*

Lemma 9.4 *If the metric space M is complete and doubling, then $WT(M)$ is closed as a collection of pointed metric spaces.*

The proof of this lemma is tedious but straightforward. It is helpful to start with a fixed embedding of $(M, d(x,y)^\alpha)$ into $\mathbf{R}^n$ for some choices of α, n, as discussed in the previous section. In fact one can use this embedding to practically reduce to the case of subsets of a fixed Euclidean space. Given a convergent sequence in $WT(M)$ one can approximate it well by a sequence of $M_{p,t}$'s, and then show that the limit of the original sequence can be realized as a limit of $M_{p,t}$'s. We omit the details.

Lemma 9.5 *Suppose that M is a complete metric space which is doubling and that $N \in WT(M)$. Then $WT(N) \subseteq WT(M)$.*

This is not hard to prove directly, but it is a little easier to use the preceding lemma to reduce to showing simply that $N_{q,r} \in WT(M)$ whenever $N \in WT(M)$, $q \in N$, and $0 < r \leq \operatorname{diam} N$. We omit the details.

Next we consider Ahlfors regularity of weak tangents.

Definition 9.6 (Ahlfors regularity for families) *A collection $\mathcal{E}$ of metric spaces is said to be Ahlfors regular if each element of $\mathcal{E}$ is Ahlfors regular with the same dimension and a uniformly bounded constant.*

Lemma 9.7 *If M is an Ahlfors regular metric space, then $WT(M)$ is an Ahlfors regular family of the same dimension.*

This follows from Lemma 8.29. One has only to observe that if M is regular, then so is $M_{p,t}$, and with the same constant.

9.3 Limits of BPI spaces

Proposition 9.8 *Suppose that $\{(M_j, d_j(x,y), p_j)\}$ is a sequence of pointed metric spaces which converges to the pointed metric space $(M, d(x,y), p)$ as in Definition 8.9, and that the $(M_j, d_j(x,y))$'s are all BPI, with the same dimension d, and with uniform choices of BPI and Ahlfors regularity constants. Then $(M, d(x,y))$ is BPI, with bounds on the BPI and Ahlfors regularity constants in terms of the bounds for the $(M_j, d_j(x,y))$'s.*

If moreover the $(M_j, d_j(x,y))$'s are all BPI equivalent, with bounded constants for the BPI equivalence, then $(M, d(x,y))$ is BPI equivalent to the $(M_j, d_j(x,y))$'s, with constants controlled by the ones in our assumptions.

Corollary 9.9 (Weak tangents of BPI spaces) *If $(M, d(x,y))$ is BPI, then every element of $WT(M)$ is BPI and BPI equivalent to M, with uniformly bounded constants.*

The corollary is an immediate consequence of the proposition. One has only to notice that if M is BPI, then the $M_{p,t}$'s are BPI and BPI equivalent to each other, all with bounded constants. This follows easily from the definitions.

Let us prove now the proposition. Let $\{(M_j, d_j(x,y), p_j)\}$, etc., be as above. We know already from Lemma 8.29 that $(M, d(x,y))$ is Ahlfors regular, with suitable estimates. Let $x, y \in M$ and $0 < r, t \leq \operatorname{diam} M$ be given. We need to find a closed subset A of $B_M(x,r)$ of substantial size which admits a conformally bilipschitz mapping in $B_M(y,t)$ with uniform bounds. We shall obtain these as limits of their counterparts from the M_j's.

Our convergence assumption implies there is an $\alpha \in (0,1]$ and a family of bilipshitz embeddings $f_j : (M_j, d_j(\cdot,\cdot)^\alpha) \to (\mathbf{R}^n, |x-y|)$, $f : (M, d(\cdot,\cdot)^\alpha) \to (\mathbf{R}^n, |x-y|)$ (with bounded bilipschitz constants) such that $f_j(p_j) = 0$ and $f(p) = 0$ for all j and $f_j(M_j)$ converges to $f(M)$ as closed subsets of $\mathbf{R}^n$. The definition of convergence ensures that there exist points $x_j, y_j \in M_j$ such that

$$\lim_{j\to\infty} f_j(x_j) = x, \qquad \lim_{j\to\infty} f_j(y_j) = y. \tag{9.2}$$

Since the M_j's are all BPI we can find closed subsets A_j of $B_{M_j}(x_j, r)$ with $H^d(A_j) \geq \theta\, r^d$ for all j and K-conformally bilipschitz mappings $\phi_j : A_j \to B_{M_j}(y,t)$ with scale factor t/r, where θ and K are controlled in terms of our bounds on the BPI constants. (Actually we are cheating slightly here. To apply the BPI condition we ought to know that $r, t \leq \operatorname{diam} M_j$ for all j, and that might not be true for finite values of j, but only in the limit. It is easy to fudge the radii slightly to make this work, at least for large j.)

It is easy to see that the sets $f_j(A_j)$ lie in a bounded subset of $\mathbf{R}^n$, and so Lemma 8.2 applies to say that we may assume that the sets $f_j(A_j)$ converge in $\mathbf{R}^n$, after passing to a subsequence if necessary. The limit is a compact subset of $f(M)$, and so it has the form $f(A)$ for some compact set $A \subseteq M$. In fact $A \subseteq \overline{B}_M(x, r)$, as one can derive from the corresponding property of the A_j's and the convergence of the metrics on the M_j's to the metric on M.

Set $\Phi_j = f_j \circ \phi_j \circ f_j^{-1}$. These are mappings from $f_j(A_j)$ into $f_j(\overline{B}_{M_j}(y, t))$. The latter are also contained in a uniformly bounded subset of $\mathbf{R}^n$, and we know from our assumptions that the Φ_j's are bilipschitz with a constant bounded independently of j (depending on t/r, but we do not care for the moment). Using Lemma 8.6 we may pass to a subsequence to conclude that the Φ_j's converge to a mapping $\Phi : f(A) \to \mathbf{R}^n$. In fact Φ takes values in $f(M)$, and so we can write it as $\Phi = f \circ \phi \circ f^{-1}$ for some mapping $\phi : A \to M$. Again one can use the convergence of the metrics to conclude that ϕ actually takes values in $\overline{B}_M(y, t)$. Also, the convergence of the metrics permits us to conclude that ϕ is K-conformally bilipschitz with scale factor t/r.

We need to have a lower bound on $H^d(A)$, namely that it is not too small compared to r^d. This we get from Lemma 8.35. We can apply this lemma because A_j converges to A in the sense of Definition 8.30 by construction.

This proves that M is BPI. Suppose now that the M_j's are BPI equivalent with uniformly bounded constant, and we want to show that M is BPI equivalent to them. It suffices to show that M is BPI equivalent to M_1, by Proposition 7.5. Let $x_1 \in M_1$, $0 < r_1 \leq \operatorname{diam} M_1$, $y \in M$, and $0 < t \leq \operatorname{diam} M$ be given. Choose $y_j \in M_j$ so that $\lim_{j\to\infty} f_j(y_j) = f(y)$, as before. For each j we can find a compact set $E_j \subseteq B_{M_1}(x_1, r_1)$ with $H^d(E_j) \geq \delta$ and a K-conformally bilipschitz mapping $g_j : E_j \to B_{M_j}(y_j, t)$ with scale factor t/r_1, with $K, \delta > 0$ controlled by the constants of BPI equivalence. Again we are cheating slightly here, because $t \leq \operatorname{diam} M_j$ may fail to hold, but it is true in the limit, so we can always correct the lie by fudging the radii slightly and restricting ourselves to large j. Again we want to take limits to get a set $E \subseteq \overline{B}_{M_1}(x_1, r_1)$ and a K-conformally bilipschitz mapping $g : E \to B_M(y, t)$ with scale factor t/r_1. The argument is practically the same as before, a little easier even. After passing to a subsequence we can assume that the E_j's do converge to a subset E of $\overline{B}_{M_1}(x_1, r_1)$. For this one can even use ordinary Hausdorff convergence, since they lie in a fixed metric space, but we can be consistent with the other argument and use convergence of the $f_j(E_j)$'s in $\mathbf{R}^n$, etc. We can then pass to a subsequence again to get a limit of the g_j's, still conformally bilipschitz with the same bounds. The remaining point is that we have a good lower bound on $H^d(E)$, and that we have for the same reason as before. (At bottom the argument is even a little easier, since the sets lie in a fixed space.)

This proves Proposition 9.8. Note that the two parts of the argument were the same because they are really special cases of one more general fact, where one decouples the domain and range. We shall not need this, but one should notice its existence.

9.4 Weak tangents of subsets

In the next sections we shall see some examples of compactness arguments for BPI spaces, but before we get to that we should address a small technical point: what happens if we take a weak tangent of a subset of a metric space? For the record let us state a trivial observation.

Lemma 9.10 *Let M be a metric space which is doubling and complete, let $\{p_j\}$ be a sequence of points in M, let $\{t_j\}$ be a sequence of positive numbers such that $t_j \le \operatorname{diam} M$ for all j, and let $\{A^j\}$ be a sequence of closed subsets of M such that $p_j \in A_j$ for all j. Define M_{p_j,t_j} and $A^j_{p_j,t_j}$ as in Section 9.1. Then after passing to a subsequence we may assume that M_{p_j,t_j} converges to a pointed metric space T and $A^j_{p_j,t_j}$ converges to a pointed metric space S, where S admits an isometric embedding into T which preserves the basepoints.*

Note that the definition of $A^j_{p_j,t_j}$ here is potentially illegal for the definition of weak tangents, because we might have that the $t_j > \operatorname{diam} A_j$. This does not bother us for the moment, it just means that in the limit the $A^j_{p_j,t_j}$'s might shrink to a point.

The proof of the lemma is a straightforward consequence of the definitions. One starts with an embedding of M into some $\mathbf{R}^n$ of the usual type, this gives embeddings for the M_{p_j,t_j} and the $A^j_{p_j,t_j}$. We pass to subsequences to get the existence of the relevant limits, and the inclusions of the A^j's into M give rise to the limiting isometry. (We could also quote our earlier results, such as Lemma 8.20 to get the isometry, since 1-bilipschitz mappings are isometries, but it is easier to understand what is happening by going back to the definitions.)

Next we want to record some criteria for knowing that the weak tangents of subsets coincide with the weak tangent of the larger space. We begin the case where we blow up a set at a single point.

Definition 9.11 (Point of thickness) *Let $(M, d(x,y))$ be a metric space, and let E be a subset of M. We say that $p \in M$ is a* point of thickness *for E if $p \in E$ and if for each $\epsilon > 0$ there is a $\delta > 0$ such that if $0 < t < \delta$ and $x \in B(p,t)$ then* $\operatorname{dist}(x, E) < \epsilon t$.

Lemma 9.12 *Suppose that $(M, d(x,y))$ is a metric space which is doubling and complete, and let E be a closed subset of M. Assume that p is a point of thickness for E. If $\{t_j\}$ is a sequence of positive numbers with $t_j \to 0$, and if $\{M_{p,t_j}\}$ converges as a sequence of pointed metric spaces (see Section 9.1), then $\{E_{p,t_j}\}$ also converges, and to an isometrically equivalent limit.*

This is easy to derive from the definitions. If we have embeddings for the M_{p,t_j}'s with the correct properties, then we still have the right properties when we restrict ourselves to the E_{p,t_j}'s, and thus we have a natural way to include the limit of the E_{p,t_j}'s in the limit of the M_{p,t_j}'s. The only point then is to observe that this inclusion becomes a surjection in the limit. This is easy to check. (Put another way, this notion of point of thickness is preserved by the kind of embeddings used in Definition 8.9.)

As a practical matter we can often verify the point of thickness using the following.

Lemma 9.13 *If $(M, d(x,y))$ is Ahlfors regular, E is a closed subset of M, and $p \in E$ is a point of density of E, then p is a point of thickness of E.*

This is easy to check, using the definitions.

Next we consider the more general situation where we take weak tangents by following a sequence of balls rather than blowing up at a single point. It will be convenient to settle for slightly less in this case.

Definition 9.14 (Asymptotic thickness) *Let $(M, d(x,y))$ be a metric space, let $\{p_j\}$ be a sequence of points in M, and let $\{t_j\}$ be a sequence of positive numbers with $t_j \leq \operatorname{diam} M$ for all j. Also let $\{A^j\}$ be a sequence of subsets of M with $A^j \subseteq \overline{B}(p_j, t_j)$ and $p_j \in A^j$ for all j. We say that $\{A^j\}$ is* asymptotically thick *in the sequence of balls $\{B(p_j, t_j)\}$ if for each $\epsilon > 0$ there is a $K > 0$ such that*

$$\operatorname{dist}(x, A^j) < \epsilon\, t_j \qquad \textit{whenever } x \in B(p_j, t_j) \textit{ and } j > K. \tag{9.3}$$

Lemma 9.15 *Suppose that $(M, d(x,y))$ is a metric space which is doubling and complete. Let $\{p_j\}$ be a sequence of points in M, let $\{t_j\}$ be a sequence of positive numbers with $t_j \leq \operatorname{diam} M$ for all j, and let $\{A^j\}$ be a sequence of closed subsets of M with $A^j \subseteq \overline{B}(p_j, t_j)$ and $p_j \in A^j$ for all j. Assume that $\{A^j\}$ is asymptotically thick in the sequence of balls $\{B(p_j, t_j)\}$. Then we can pass to a subsequence in such a way that the sequences $\{M_{p_j,t_j}\}$ and $\{A^j_{p_j,t_j}\}$ converge to pointed metric spaces T and S, respectively, and there is an isometric embedding of S into T which respects the basepoints and maps $\overline{B}_S(q,r)$ onto $\overline{B}_T(p,r)$ for each $0 < r < 1$.*

In this case we do not get that the weak tangents coincide entirely, just on a large piece, but that will fine in practice.

This lemma is again easy to derive from the definitions. We start with an embedding for M into some $\mathbf{R}^n$ of the usual type, that gives us embeddings for the M_{p_j,t_j}'s and the $A^j_{p_j,t_j}$'s. We pass to subsequences to get convergence for both sequences. In the limit there is a natural inclusion of S into T, because of the inclusions of the A_{p_j,t_j}'s in the M_{p_j,t_j}'s for each j. It is easy to check that the inclusion is an isometry, and that the limiting map is surjective on balls of radius < 1 because of the thickness hypothesis. (Indeed this is the only place where we use the thickness hypothesis.)

There is a useful criterion for the existence of asymptotic thickness analogous to Lemma 9.13, as follows.

Lemma 9.16 *Let M be a metric space and let A be a nonempty subset of M. Then either A is porous (Definition 5.7), or there exist a sequence $\{p_j\}$ of points in A and a sequence $\{t_j\}$ of positive real numbers with $t_j \leq \operatorname{diam} A$ for all j such* *that $\{A \cap \overline{B}(p_j, t_j)\}$ is asymptotically thick in the sequence of balls $\{B(p_j, t_j)\}$. If A is contained in a compact subset of M, then we may also require that $\lim_{j\to\infty} t_j = 0$.*

The first part of the lemma follows straight from the definition of porosity. As for the second, suppose that A is contained in a compact subset of M but the t_j's do not tend to zero. By passing to a subsequence we may assume that the t_j's are bounded away from 0 and that the p_j's converge to a point $p \in M$. In this case we get that the closure of A contains a neighborhood of p, and then there is no trouble in replacing the t_j's so that they tend to 0.

Note that if M is Ahlfors regular then Lemma 5.8 applies to control the dimension of bounded subsets of A when A is porous.

9.5 Comparisons with rectifiability

It is well known that rectifiable sets in Euclidean spaces can be characterized in terms of the existence of approximate tangent planes at almost all points. (See [Fe, Ma].) A self-similarity condition like BPI implies that the given set looks practically the same everywhere, and so it should be enough to know the asymptotic behavior of the set at one point to know the set almost completely.

This idea can be made precise in various ways, and we confine ourselves to a simple one here. Let us work with a set E in $\mathbf{R}^n$ rather than an abstract metric space, a subset which is Ahlfors regular of dimension d. Given $x \in E$ and $t > 0$ set

$$\beta(x,t) = \inf_P \sup_{y \in E \cap B(x,t)} \frac{\operatorname{dist}(y,P)}{t}. \tag{9.4}$$

The infimum here is taken over all d-planes in $\mathbf{R}^n$. This quantity measures the extent to which E nearly lives on a d-plane inside $B(x,t)$. It is normalized to be scale invariant, and so that $\beta(x,t) \leq 1$ always by definition.

Proposition 9.17 *Suppose that E is a d-dimensional regular set in $\mathbf{R}^n$ for which there exists a sequence of points $\{p_j\}$ in E and a sequence $\{t_j\}$ of positive numbers with $t_j \leq \operatorname{diam} E$ for all j and*

$$\lim_{j\to\infty} \beta(p_j, t_j) = 0. \tag{9.5}$$

Then there is a weak tangent to E which is isometrically equivalent to a subset of $\mathbf{R}^d$.

If E is also BPI, then E is uniformly rectifiable (with respect to $\mathbf{R}^d$).

This is easy to prove. Let E, $\{p_j\}$, $\{t_j\}$ be given as above. We observe first that we can strengthen our hypothesis slightly, to get a sequence $\{r_j\}$ of positive numbers such that $r_j \leq t_j$ for all j and

$$\lim_{j\to\infty} \frac{t_j}{r_j} = \infty, \qquad \lim_{j\to\infty} \frac{t_j}{r_j}\, \beta(p_j, t_j) = 0. \tag{9.6}$$

This is easy to arrange, and the r_j's are more convenient radii to use for passing to a weak tangent.

Since we are working inside a Euclidean space, we may as well take E_{p_j,r_j} to be the image of E under $z \mapsto r_j^{-1}(z - p_j)$ (rather than working with the more abstract notion for metric spaces in general). By passing to a subsequence we may suppose that E_{p_j,r_j} converges to a subset T of $\mathbf{R}^n$, and T represents the weak tangent to E corresponding to this sequence and with the origin as basepoint. One can check that (9.6) implies that T lies inside a d-plane. This implies the first part of the proposition.

Now suppose that E is a BPI set. Then T is also BPI and E is BPI equivalent to T, by Corollary 9.9. This implies that there are subsets of E which are bilipschitz equivalent to subsets of $\mathbf{R}^d$, sets with positive measure, and we even have bounds if we want them. Therefore E is BPI equivalent to $\mathbf{R}^d$, by Proposition 7.1, and uniformly rectifiable by Proposition 7.6. This completes the proof of Proposition 9.17.

To put the proposition into perspective we should recall that there are known criteria for deducing rectifiability or uniform rectifiability properties of a set in terms of something like smallness conditions on the $\beta(x,t)$'s. (We mean general sets here, not the special case of BPI sets.) We mentioned before the characterization of rectifiability in terms of the existence almost everywhere of approximate tangent planes. This is quite different from the requirement that $\lim_{t\to 0}\beta(x,t) = 0$ almost everywhere, because the $\beta(x,t)$'s permit the approximating d-plane to spin around as $t \to 0$, while the notion of approximate tangent space does not. (The concept of an approximate tangent plane does allow other deteriorations of a technical nature which are not so important for this point.) If one wants a criterion for rectifiability directly in terms of quantities like the $\beta(x,t)$'s, then one should impose stronger quadratic conditions on them, as in [J1, J3, DS2]. These stronger conditions are roughly analogous to the requirement that a sequence be square summable instead of simply tending to 0. To get a condition more like $\beta(x,t) \to 0$ one should work with "bilateral" versions of the $\beta(x,t)$'s, which measure the extent to which the set lies close to a d-plane and the d-plane lies close to the set. The $\beta(x,t)$'s do only the first and permit holes, i.e., the weak tangent T is allowed to be a proper subset of $\mathbf{R}^d$. See [Ma, DS4] for rectifiability results concerning the bilateral versions of the $\beta(x,t)$'s.

Thus conditions in the realm of $\beta(x,t) \to 0$ are not sufficient in general to ensure rectifiability. In fact we have the following.

Proposition 9.18 *There is a compact subset E of $\mathbf{R}^2$ which is regular of dimension 1 such that*

$$\lim_{t\to 0}\sup_{x\in E}\beta(x,t) = 0 \tag{9.7}$$

and such that E is totally unrectifiable, i.e., if A is any compact subset of $\mathbf{R}^2$ which is bilipschitz equivalent to a subset of the real line, then the intersection of E with A has measure 0.

See [DS2], p.135–7.

Proposition 9.19 *If E has the properties mentioned in Proposition 9.18, and if A is a closed subset of E which is bilipschitz equivalent to a subset of a BPI metric space of dimension 1, then $|A| = 0$.*

Thus the set is not only totally unrectifiable in the sense of ordinary geometric measure theory but it is also "unrectifiable" with respect to any BPI geometry. The existence of such a set is not a big surprise, but it is good to know. It is also amusing that it occurs with sets that are close to being rectifiable, in their tangential behavior, although this is actually pretty natural.

Let us prove the proposition. Let E be as above, and suppose to the contrary that there is a BPI metric space M of dimension 1, a closed subset A of E of positive measure, and a bilipschitz mapping $h : A \to M$. Then A has a point a of thickness, by Lemma 9.13. Set $p = h(a) \in M$. Let $\{t_j\}$ be a sequence of positive numbers which tends to zero. By passing to a subsequence if necessary we may suppose that $\{E_{a,t_j}\}$ converges to a space $T \in WT(E)$. By Lemma 9.12 T is also a weak tangent for A, with basepoint a. Of course we are using Lemma 8.13 here (although in this case we could just go back to Lemma 8.2). Similarly we may assume that $\{M_{p,t_j}\}$ converges to a space $W \in WT(M)$, by passing to a subsequence. We can view h as really a sequence of mappings $h : A_{a,t_j} \to M_{p,t_j}$, and as such it is bilipschitz with a uniform constant that does not depend on j (as one can easily check), and of course h respects the basepoint, $h(a) = p$ by definition. This permits us to pass to a subsequence and obtain a limiting mapping $H : T \to W$, as in Lemma 8.22. More precisely, our sequence of mappings satisfies the equicontinuity and uniform boundedness assumptions in Lemma 8.22 because of the uniform Lipschitz condition and basepoint conditions. This limiting mapping H is bilipschitz, as in Lemma 8.20.

In fact we know that T is a 1-dimensional regular subset of a line, as in (the proof of) Proposition 9.17. Thus we get a bilipschitz mapping from a nontrivial subset of the line into W. Since we know that W is BPI (Corollary 9.9) we conclude that W is BPI equivalent to the real line, as in Proposition 7.1. On the other hand M is BPI equivalent to W, by Corollary 9.9, and so M is BPI equivalent to the real line by Proposition 7.5. Proposition 7.4 implies then that M is covered by a countable union of bilipschitz images of subsets of the real line together with a set of measure 0. Our original assumption that E has a subset of positive measure which is bilipschitz equivalent to a subset of M implies now that E has a subset of positive measure which is bilipschitz equivalent to a subset of the real line. This contradicts the unrectifiability assumption on E, and the proposition follows.

Incidentally, if we wanted we could have chosen the point a a little more carefully so that $h(a)$ is a point of density of $h(A)$, and hence a point of thickness. This would imply that the mapping H was actually a bilipschitz mapping of T onto W.

One might interpret the proposition as saying that there are regular sets which have nothing in common with any BPI geometry. This is partially true but not completely true, since by assumption these sets are asymptotically Euclidean at all points. It is not clear to what extent one can always find some vestige of BPI geometry within a given regular set. There are plenty of negative statements of this nature, but it would be nice if one could find a positive statement.

9.6 BPI spaces which are not BPI equivalent

We know from Proposition 7.1 that if M and N are two BPI spaces which are not BPI equivalent, then any pair of subsets of M and N which are bilipschitz equivalent must have measure 0. This says that the two spaces have nothing to do with each other in a certain sense, but Proposition 9.20 below provides a much stronger version of this distinction. The main point is that two BPI spaces that are different must also be different "asymptotically", because of Corollary 9.9. This should be compared with the set provided by Proposition 9.18, which is unrectifiable and therefore distinct from Euclidean geometry at all definite scales but then asymptotically Euclidean at all points in a uniform way.

Proposition 9.20 *Let M and N be BPI metric spaces of dimension d which are not BPI equivalent, and suppose that A is a subset of M which is bilipschitz equivalent to a subset of N. Then A is porous. In particular there is an $\eta > 0$ so that A is semi-regular of dimension $d - \eta$ (by Lemma 5.8).*

In fact we have the following result, which is slightly stronger.

Proposition 9.21 *Let M and N be BPI metric spaces of dimension d which are not BPI equivalent. For every $K > 0$ there is an $\epsilon > 0$ so that if $x \in M$, $0 < r \leq \operatorname{diam} M$, $A \subseteq B_M(x, r)$, and $\phi : A \to N$ is K-conformally bilipschitz, then there is a point $y \in B_M(x, r/2)$ such that $\operatorname{dist}(y, A) \geq \epsilon\, r$.*

It is easy to see that Proposition 9.21 implies Proposition 9.20, by definition of porous sets (Definition 5.7).

Let us now prove Proposition 9.21. Let M, N, and K be given. We argue by contradiction and use compactness. If the proposition is false, then we can find a sequence of points $\{x_j\}$ in M, a sequence of radii $\{r_j\}$ with $0 < r_j \leq \operatorname{diam} M$ for all j, a sequence of sets $A^j \subseteq \overline{B}_M(x_j, r_j)$, and a sequence of K-conformally bilipschitz mappings $\phi_j : A^j \to N$ such that

$$\operatorname{dist}(y, A^j) \leq \epsilon_j\, r_j \qquad \text{when } y \in B_M(x_j, r_j/2) \tag{9.8}$$

for all j, and such that $\epsilon_j \to 0$ as $j \to \infty$. We may as well require that the A^j's all be closed.

For each j choose a point $a_j \in A^j$, set $b_j = \phi_j(a_j)$, and set $t_j = \operatorname{diam} \phi_j(A^j)$. Consider the pointed metric spaces M_{a_j, r_j}, $A^j_{a_j, r_j}$, and N_{b_j, t_j}, as in Section 9.1. By passing to a subsequence we may assume that these spaces converge to the pointed metric spaces T, S, and W. From Lemma 9.10 we have that there is an isometric embedding of S into T which respects the basepoints (and which merely reflects the fact that the A^j's are subsets of M).

Consider now the mappings $\phi_j : A^j_{a_j, r_j} \to N_{b_j, t_j}$. These mappings preserve the basepoints by definition, and one can check that they are uniformly bilipschitz.

This uses the fact that the ϕ_j's are K-conformally bilipschitz as maps from A^j to N, and also the fact that (9.8) implies that $\operatorname{diam} A^j \geq C^{-1} r_j$ for some constant C. (Amusingly enough this constant C depends on the regularity constant for M. The point of this bound is that the scale factors for the K-conformally bilipschitz mappings $\phi_j : A^j \to N$ must be comparable to $\operatorname{diam} \phi_j(A_j) / \operatorname{diam} A_j$, and when one works out the normalizations this translates into a uniform bilipschitz bound for the mappings $\phi_j : A^j_{a_j,r_j} \to N_{b_j,t_j}$.) In particular the mappings $\phi_j : A^j_{a_j,r_j} \to N_{b_j,t_j}$ are uniformly Lipschitz, and hence equicontinuous and uniformly bounded on bounded subsets. Therefore we may pass to a subsequence and get a limiting mapping $\Phi : S \to W$, as in Lemma 8.22. This limiting mapping is bilipschitz, because of Lemma 8.20.

There is an isometric equivalence between the open balls of radius 1/2 of the basepoints in S and T. This follows from our assumption (9.8); unfortunately this assertion does not quite come under the purview of Lemma 9.15, but the principle is the same, and it is easy to check.

The conclusion of all this is that we now have a bilipschitz embedding from a ball in T into W. From Corollary 9.9 we know that T and W are BPI and BPI equivalent to M and N, respectively. Proposition 7.1 implies now that T and W are BPI equivalent, and so we conclude that M and N are BPI equivalent, by Proposition 7.5. This contradicts our hypotheses, and Proposition 9.21 follows.

9.7 Weak tangents of mappings

Let $(M, d(x,y))$ and $(N, \rho(u,v))$ be metric spaces which are doubling and complete, and let $f : M \to N$ be a Lipschitz mapping. Given $p \in M$ and $t > 0$ set $q = f(p) \in N$ and consider f as mapping from $M_{p,t}$ into $N_{q,t}$, where the latter are defined as in Section 9.1. To be compatible with the definition given there we should demand also that $t \leq \min(\operatorname{diam} M, \operatorname{diam} N)$. This assumption is not a big deal, we just want to prevent trivial limits, where spaces collapse down to points.

A *weak tangent* to f is a mapping $g : A \to E$ with the following properties. A is a weak tangent to M, and more precisely A is a pointed metric space with basepoint a which arises as the limit of M_{p_j,t_j}, where each p_j is an element of M and $\{t_j\}$ is a sequence of numbers such that

$$0 < t_j \leq \min(\operatorname{diam} M, \operatorname{diam} N) \qquad \text{for all } j. \tag{9.9}$$

Similarly E is a weak tangent to N, E is a pointed metric space with basepoint e, and we ask that E arise as the limit of N_{q_j,t_j}, where $q_j = f(p_j)$ and we use the same t_j's as for A. Finally we ask that g be the limit of the mappings $f : M_{p_j,t_j} \to N_{q_j,t_j}$ (where we view these copies of f as being different because of the different basepoints and metrics). In particular we require that $g(a) = e$. To be precise we use here the notion of convergence given in Definition 8.18.

A key point here is that for any sequence $\{p_j\}$ in M and any sequence $\{t_j\}$ of numbers in between 0 and $\min(\operatorname{diam} M, \operatorname{diam} N)$ we have a subsequence for which the limit exists. This follows from Lemma 8.22. For this assertion we are using the assumption that f is Lipschitz to know that each copy $f : M_{p,t} \to N_{q,t}$ is Lipschitz with the same norm, and the basepoint condition $f(p) = q$ to know that f is uniformly bounded on bounded subsets. In other words, the equicontinuity and boundedness assumptions in Lemma 8.22 are automatic here.

We write $WT(f)$ for the collection of all weak tangents of f. Sometimes we may write $WT(M, N, f)$ to make explicit the dependence on the metric spaces M and N, and strictly speaking we ought to include the domain and range of the tangent mappings in the notation. We write $WT_p(f)$ for the weak tangents which arise through limits with $p_j = p$ for all j and $t_j \to 0$ as $j \to \infty$, i.e., the weak tangents obtained by blowing up at the point p.

Note that if f is L-Lipschitz, then so are all of its weak tangents. Similarly, if f is L-bilipschitz, then so are all of its weak tangents. See Lemma 8.20.

As for the concept of weak tangents of spaces, it does not really matter which choices of embeddings, etc., of the $M_{p,t}$'s and $N_{q,t}$'s into Euclidean spaces that we use to define this notion of weak tangents (through Definition 8.18). We might as well start with any embedding for M and N (satisfying the usual bilipschitz conditions), then use those embeddings to get embeddings for the $M_{p,t}$'s and $N_{q,t}$'s, and then use only these embeddings. As before, the existence and uniqueness results for convergence of mapping packages ensure that any sequential limit that would appear for arbitrary embeddings will also appear for these prescribed embeddings, up to isometric equivalence. We are using here the fact that the definition of weak tangents permits the passage to subsequential limits freely.

As in Lemma 9.5, we have the following.

Lemma 9.22 *$WT(g) \subseteq WT(f)$ whenever $g \in WT(f)$.*

The proof is tedious but not surprising, and we omit the details. More generally, limits of weak tangents are weak tangents, as before.

Note that we really encountered the idea of weak tangents of mappings already, in Section 9.5 (especially the proof of Proposition 9.19) and in Section 9.6. Technically there were some small differences between the definition given here and the constructions used there, but the basic principles are the same.

9.8 Weak tangents of measures

Let $(M, d(x, y))$ be a metric space which is doubling and complete, and suppose that μ is an Ahlfors regular measure on M of dimension d (Definition 8.27). Given $p \in M$ and $t > 0$ set $\mu_{p,t} = t^{-d}\mu$, viewed as a measure on $M_{p,t}$. Thus $\mu_{p,t}$ does not really depend on p, but it is helpful to keep track of the basepoints. With respect to our usual metric on $M_{p,t}$ we have that $\mu_{p,t}$ is Ahlfors regular of dimension d and with a uniformly bounded constant that does not depend on p or t.

By a *weak tangent* to μ we mean a nonnegative Borel measure ν on a metric space N which is complete and doubling such that N is a weak tangent to M, arising as the limit of M_{p_j,t_j} for some sequences $\{p_j\}$, $\{t_j\}$, and such that ν is the limit of μ_{p_j,t_j} on M_{p_j,t_j} in the sense of Definition 8.25. We also demand that ν be nonzero, which is equivalent to requiring that that the distance between p_j and the support of μ in M is bounded by a constant multiple of t_j.

Let us record some basic facts about weak tangents of measures. Given any pair of sequences $\{p_j\}$, $\{t_j\}$ which satisfy our auxiliary conditions ($0 < t_j \leq \operatorname{diam} M$ for all j, and the distance from p_j to the support of μ is $O(t_j)$), there is a subsequence for which the limit exists. For this we use Lemma 8.26. Every weak tangent of μ is also Ahlfors regular of dimension d, because of Lemma 8.28.

If the support of μ is all of M, then the support of any weak tangent ν is all of the corresponding metric space N. This follows from Lemma 8.28.

We write $WT(\mu)$ for the collection of all weak tangents to μ, or $WT(M, \mu)$ if we want to be explicit about the dependence on M. If $\nu \in WT(\mu)$, then $WT(\nu) \subseteq WT(\mu)$. To be more precise, we should treat weak tangents as ordered pairs of spaces and measures, and then say that if $(N, \nu) \in WT(M, \mu)$, then $WT(N, \nu) \subseteq WT(M, \mu)$.

More generally, limits of weak tangents are weak tangents.

We have restricted ourselves to Ahlfors regular measures here because it is convenient to do so and sufficient for our purposes, but one can make some easy extensions.

See [Ma] for another view of tangent measures.

10

REST STOP

So far we have the concepts of BPI spaces and BPI equivalence and some basic properties of them. In Chapters 6 and 7 we saw how the BPI condition and BPI equivalence are related to various other conditions. The techniques were based on coverings and cutting spaces up into little pieces. In Chapters 8 and 9 we established the basic machinery of compactness.

In short we have some basic tools. We also have some examples.

So where do we go from here?

Our next goal is to be able to compare BPI spaces when they are not BPI equivalent. For this purpose it is natural to look at mappings between BPI spaces, for instance. We shall also look at more examples.

We shall be finding ourselves much more in the situation where we have some amusing concepts but not the kinds of theorems that we want. We have a lot of counterexamples to the more optimistic conjectures, but basic questions remain unanswered.

11

SPACES LOOKING DOWN ON OTHER SPACES

11.1 Definitions and basic facts

Definition 11.1 (Looking down) *Let M and N be two BPI spaces of the same dimension. We say that M* looks down *on N if there is a closed subset A of M and a Lipschitz mapping $f : A \to N$ such that $f(A)$ has positive measure.*

This should be compared with BPI equivalence, or more precisely Proposition 7.1, in which we would ask that f be bilipschitz. Now we are trying to split BPI equivalence into two pieces, so to speak.

Problem 11.2 *If M and N are BPI spaces of the same dimension such that each looks down on the other, then is it true that M and N are BPI equivalent?*

If this were true it would be very nice. As we shall see, it would mean that "looking down" is a partial ordering on the collection of BPI equivalence classes. We have counterexamples to some statements that would imply this, which we shall discuss later, but we do not have a clear view of the question itself.

To compensate for this uncertainty we introduce another definition.

Definition 11.3 (Look-down equivalence) *Two BPI spaces of the same dimension are said to be* look-down equivalent *if each looks down on the other.*

Of course BPI equivalence implies look-down equivalence.

Let us establish now some simple properties of the looking-down relation.

Proposition 11.4 *Let M and N be two BPI spaces of the same dimension such that M looks down on N. Then for every measurable subset E of M with positive measure we can find a sequence of measurable subsets $\{E_i\}$ of E and a sequence of Lipschitz mappings $f_i : E_i \to N$ so that $N \setminus \bigcup_i f_i(E_i)$ has measure zero.*

This is an easy consequence of the definition and Proposition 6.11.

Proposition 11.5 (Transitivity) *Let M, N, and P be BPI spaces of the same dimension, and assume that M looks down on N and that N looks down on P. Then M looks down on P.*

This follows easily from Proposition 11.4.

Proposition 11.6 *If two BPI spaces of the same dimension are look-down equivalent, then they look down on the same BPI spaces.*

This is an immediate consequence of the preceding proposition.

These are the basic facts about the look-down relation. In the next section we address the nuisance of mappings not being defined everywhere, and afterwards we discuss examples.

11.2 Mappings defined everywhere

It is slightly unpleasant that the mapping f in the definition of looking down is not required to be defined everywhere. Let us ameliorate this deficiency with the following.

Lemma 11.7 *Let M and N be two metric spaces which are Ahlfors regular of the same dimension d and also BPI, and assume that M looks down on N. Then there exist spaces $M' \in WT(M)$ and $N' \in WT(N)$ and a Lipschitz mapping $g : M' \to N'$ such that $g(M')$ has positive measure.*

We shall obtain stronger results in Chapter 12.

To prove the lemma we make a blowing up. Let A be a closed subset of M and $f : A \to N$ a Lipschitz mapping such that $f(A)$ has positive measure, as in Definition 11.1. We may as well assume that A is compact.

Sublemma 11.8 *There exists $x \in M$ such that x is a point of density of A and*

$$\liminf_{t \to 0} t^{-d} H^d(f(A \cap B_M(x,t))) > 0. \tag{11.1}$$

Suppose not. We want to contradict the assumption that the image of f has positive measure.

Let $\epsilon > 0$ be given, and let A_1 denote the set of points of density in A. If $x \in A_1$ then our assumption implies that (11.1) fails, and hence there exists $t \in (0,1)$ such that

$$H^d(B_M(x,t) \cap A) \geq C^{-1} t^d \tag{11.2}$$

but

$$H^d(f(B_M(x,5t) \cap A)) \leq \epsilon\, t^d. \tag{11.3}$$

This constant C depends only on the Ahlfors regularity constant for M and nothing else. (We get it by taking t small enough to be able to use the point of density condition.) Let $\mathcal{B}$ denote the collection of all balls $B_M(x,t))$ obtained in this manner. As in Section 5.2 there is a sequence of pairwise disjoint balls $\{B_i\}$ in $\mathcal{B}$ such that either

$$A_1 \subseteq \bigcup_i 5B_i \tag{11.4}$$

or there are infinitely many B_i's and their radii remain bounded away from 0. However, since the B_i's are pairwise disjoint we have that

$$\sum_i (\text{radius}\, B_i)^d \leq C \sum_i H^d(B_i \cap A) \leq C\, H^d(A), \tag{11.5}$$

using also (11.2). Our requirement that A be compact ensures that this sum converges, so that we cannot have infinitely many B_i's with radii bounded from below. Thus we have (11.4). From (11.3) we obtain that

$$\begin{aligned} H^d(f(A_1)) &\leq \sum_i H^d(f(A_1 \cap 5B_i)) \\ &\leq \sum_i \epsilon \, (\text{radius}\, B_i)^d \leq C \, \epsilon \, H^d(A). \end{aligned} \tag{11.6}$$

This constant C does not depend on ϵ, and we conclude that $H^d(f(A_1)) = 0$, since ϵ was arbitrary. We have $H^d(f(A \backslash A_1)) = 0$ automatically, because $H^d(A \backslash A_1) = 0$, and hence we obtain $H^d(f(A)) = 0$, in contradiction to our assumptions. This proves the sublemma.

Let us now derive the lemma from the sublemma. Choose x as in the sublemma, and let $\{t_j\}$ be a sequence of positive numbers which tends to 0. Set $y = f(x)$, a point in N, and consider now f as a mapping from M_{x,t_j} into N_{y,t_j}. We are using here the notation of Section 9.1. We want to blow up f at x along $\{t_j\}$, except that we are happy to pass to subsequences whenever necessary. By passing to subsequences we may assume that M_{x,t_j} converges to a pointed metric space $M' \in WT(M)$ and that N_{y,t_j} converges to a pointed metric space $N' \in WT(N)$, as in Lemma 8.13. Because x is a point of density of A, it is a point of thickness (Definition 9.11, Lemma 9.13), and A_{x,t_j} also converges to M', as in Lemma 9.12.

Because of Lemma 8.22 we may assume that the sequence of mappings $f : A_{x,t_j} \to N_{y,t_j}$ converges to a mapping $g : M' \to N'$. These mappings $f : A_{x,t_j} \to N_{y,t_j}$ are all the same in terms of their actions on points, but we are deforming the metrics. We are deforming them the same way in domain and image, in such a way that $f : A_{x,t_j} \to N_{y,t_j}$ has uniformly bounded Lipschtz constant. We have also chosen the basepoints to match up, ensuring that our sequence of mappings is uniformly bounded and equicontinuous on bounded sets, as required in Lemma 8.22.

Set $K_j = A \cap \overline{B}_M(x, t_j)$. As a subset of A_{x,t_j} this is just the closed unit ball about the basepoint. By passing to another subsequence we may assume that $\{K_j\}$ converges to a compact subset K of M'. From Lemma 8.32 we conclude that $f(K_j)$ in N_j converges to $g(K)$ in N'.

The last point now is that the H^d-measures of $f(K_j)$ in N_j in the metric of N_j – obtained by dividing the metric on N by t_j – are bounded from below by a positive number. This follows from (11.1). We conclude that $g(K)$ has positive H^d measure in N', because of Lemma 8.35.

This completes the proof of Lemma 11.7.

11.3 Cantor sets to Euclidean spaces

Set $F = \{0, 1\}$, and let F^∞ denote the set of binary sequences $x = \{x_j\}_{j=1}^\infty$. As in Section 2.3, define a metric $d(x, y)$ on F^∞ by

$$d(x, y) = 2^{-L(x,y)}, \tag{11.7}$$

where $L(x, y)$ is the largest integer l such that $x_i = y_i$ when $1 \leq i \leq l$, $L(x, y) = \infty$ when $x = y$. This defines a metric space which is Ahlfors regular of dimension 1 and which is also BPI.

There is a natural mapping $\phi : F^\infty \to [0, 1]$ given by

$$\phi(x) = \sum_{j=1}^{\infty} x_j \, 2^{-j}. \tag{11.8}$$

This mapping simply takes a binary sequence and associates to it the real number with that binary expansion. In particular it is surjective. It is not very difficult to see that this mapping is Lipschitz. Thus F^∞ looks down on $[0, 1]$.

Of course one can make many variations on this theme. One can use Cantor sets starting from a finite set with any number of elements (≥ 2), adjusting the metric and mapping accordingly. One can also take products, so as to have Cantor sets look down on cubes of higher dimension.

One might hope that a Cantor set would look down on practically anything, but they turn out to be somewhat more tricky than one might expect. We shall return to them several times.

11.4 Looking down from Euclidean spaces

Definition 11.9 (Minimal for looking down) *A BPI space M is said to be* minimal for looking down *if it has the property that if N is another BPI space of the same dimension and M looks down on N then M and N are BPI equivalent.*

Proposition 11.10 $\mathbf{R}^n$ *is minimal for looking down.*

Problem 11.11 *Which BPI spaces are minimal for looking down? Does every BPI space look down on a minimal BPI space? Is a BPI space determined up to look-down equivalence by the collection of minimal BPI spaces on which it looks down?*

To prove the proposition it suffices to know the following (because of Proposition 7.1).

Theorem 11.12 *Let M be a metric space, let A be a subset of $\mathbf{R}^n$ (compact, say), and let $f : A \to M$ be a Lipschitz mapping. If $H^n(f(A)) > 0$, then there is a subset of A of positive measure on which f is bilipschitz.*

If M were a subset of a finite dimensional Euclidean space then Proposition 11.12 would follow from standard results in analysis, for which the differentiability almost everywhere of f would be a key ingredient. The general case is more difficult, because of the absence of such a differentiability theorem for mappings into general metric spaces. One can always embed a metric space isometrically into a Banach space, and thereby have a linear structure available, but it is well known that Lipschitz mappings into Banach spaces need not be differentiable almost everywhere.

Nonetheless Theorem 11.12 turns out to be true, a result of Kirchheim [Ki]. (See Theorem 9 on p.119 of [Ki], for instance. Note that he defines rectifiability for metric spaces in the first paragraph of [Ki].)

We shall give another proof of Proposition 11.10 later, but Kirchheim's theorem gets to the heart of the general matter in a way that our argument does not.

Incidentally, there are nice open problems connected to Kirchheim's theorem, about quantitative forms of it. Here is one version.

Problem 11.13 *Let M be an Ahlfors regular metric space of dimension d. Suppose that A is a subset of the unit ball in $\mathbf{R}^d$, and that there is a Lipschitz mapping $f : A \to M$ such that $H^d(f(A)) \geq \delta$ for some $\delta > 0$. Does there exist a subset E of A such that $|E| \geq \epsilon$ and the restriction of f to E is K-bilipschitz, where $K, \epsilon > 0$ are permitted to depend on δ, d, the Lipschitz constant for f, and the regularity constants for M, but not otherwise on f or A?*

If M is a subset of some $\mathbf{R}^n$ then the answer to this question is yes, by a result of Jones [J2]. For general metric spaces it is not known. If it is true then it means that an important piece of the story of uniform rectifiability works for general metric spaces and not just for subsets of Euclidean spaces. (A lot of the rest works for other reasons. For instance, although uniformly rectifiable metric spaces need not admit bilipschitz embeddings into some $\mathbf{R}^n$, because of an example in [Se4], there is an embedding if one permits a modest deformation of the metric. This embedding result is similar to the one in [Se2]. The point is that the deformation of the metric is much milder than in Assouad's embedding theorem, and in particular does not disturb rectifiability properties.)

11.5 Euclidean and Heisenberg geometries

Fix a positive integer n, and let H_n denote the Heisenberg group, as in Chapter 4, equipped with the metric described there. Thus H_n defines a metric space which is Ahlfors regular of dimension $2n+2$. It turns out that neither of H_n or $\mathbf{R}^{2n+2}$ looks down on each other.

Suppose for instance that H_n looks down on $\mathbf{R}^{2n+2}$. This means that there is a Lipschitz mapping from a subset of H_n into $\mathbf{R}^{2n+2}$ whose image has positive measure. We can extend this to a Lipschitz mapping defined on all of H_n, as in Section 5.1.

Lipschitz mappings from H_n into $\mathbf{R}^{2n+2}$ are differentiable almost everywhere. A precise statement is given [P1], but roughly it goes as follows. In order to define a derivative at a point one needs to make a blowing up at the point. On Euclidean spaces this can be accomplished using translations and dilations, but indeed we have similar operations on the Heisenberg group even if there are some minor additional complications (like the difference between left and right translations). The existence of the derivative means first the existence of a tangent mapping in the limit of blowing up. In [P1] it is shown that not only does this limit exist almost everywhere, but it is given by a group homomorphism almost everywhere, at least if one makes the proper normalizations so that the tangent mapping is centered at the origin and takes the origin to the origin.

The Heisenberg group is not abelian. Group homomorphisms from it into the abelian Euclidean groups have to factor through the quotient by the commutator subgroup, which amounts to $\mathbf{R}^{2n}$. The bottom line is that the image of H_n under a group homomorphism into $\mathbf{R}^{2n+2}$ is a plane of dimension $\leq 2n$.

Once one knows this one can show without too much trouble that the image of the mapping has to have measure 0. The point is to use covering arguments in the standard way to reduce to the fact that the images of these tangent mappings are too small. From this we may conclude that H_n does not look down on $\mathbf{R}^{2n+2}$.

For the reverse, a convenient argument is to use Proposition 11.10 to say that if $\mathbf{R}^{2n+2}$ looks down on H_n, then the two are BPI equivalent, which would imply that H_n looks down on $\mathbf{R}^{2n+2}$, a contradiction. This argument is rather crude, using too much technology to get less than is true. It is better to argue directly, using differentiability as above, and the fact that group homomorphisms from Euclidean spaces into Heisenberg groups are strongly restricted by the nonabelian structure of the Heisenberg groups.

Thus neither of H_n or $\mathbf{R}^{2n+2}$ looks down on the other. We shall see other examples of this later.

Problem 11.14 *Given two BPI spaces M and N of the same dimension, consider the collection of equivalence classes of BPI spaces of the same dimension which look down on them both. Is there always such a BPI space? If so, what can we say about the structure of this collection? In terms of minimal elements for instance?*

Notice that two BPI spaces of the same dimension need not look down simultaneously on any BPI space. For instance, if both H_n and $\mathbf{R}^{2n+2}$ look down on some common space, then the common space must be BPI equivalent to $\mathbf{R}^{2n+2}$, by Proposition 11.10, and then H_n would look down on $\mathbf{R}^{2n+2}$, by transitivity.

It is amusing that while Euclidean space is minimal for looking down, not everything looks down on it.

Problem 11.15 *Is the Heisenberg group H_n minimal for looking down?*

This is related to the matter of making a theory like that of rectifiability for the Heisenberg group, as mentioned in Chapter 4.

Note that there are other *Carnot groups* besides the Heisenberg groups, some of which possess stronger rigidity properties. See [P1].

11.6 A Cantor set with sliding

"Sliding" here is not a technical term, it simply reflects the geometry of a concrete example based on Cantor sets.

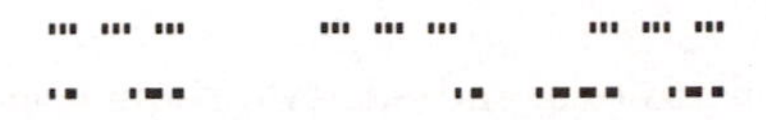

FIG. 11.1. A standard Cantor set, and one with sliding

It will be convenient to work here with Cantor sets as subsets of the real line, rather than the symbolic description that we used before. Given a closed interval I in $\mathbf{R}$, let $I(j)$, $j = 1, 2, 3, 4, 5$ denote the five closed subintervals that we get by cutting I into equal pieces and labelling them from left to right.

Let E denote the Cantor set obtained by starting with the unit interval $[0, 1]$ and repeating the rule

$$I \mapsto I(1) \cup I(3) \cup I(5). \tag{11.9}$$

That is, we start with $[0, 1]$, we replace it with three intervals according to this rule, we replace each of those with three intervals, etc. This gives us a decreasing sequence of compact sets E_j, with $E_0 = [0, 1]$ and where E_j is the union of 3^j disjoint closed intervals of length 5^{-j}, and we set $E = \bigcap_j E_j$. This set is bilipschitz equivalent to a symbolic Cantor set with three letters as described in Section 2.3.

Let F denote the set obtained from $[0, 1]$ by repeating the rule

$$I \mapsto I(1) \cup I(4) \cup I(5). \tag{11.10}$$

That is, we set $F_0 = [0, 1]$, we take F_1 to be the union of the three intervals as in the rule, and then we repeat. Although $I(4) \cup I(5)$ is actually an interval, when we repeat the rule we treat $I(4)$ and $I(5)$ as separate intervals. Thus F_j is the union of 3^j intervals of length 5^{-j}, and although these intervals are not disjoint, they have disjoint interiors.

A picture of E and F is given in Figure 11.1.

Both of these spaces are Ahlfors regular of dimension d, where d is chosen so that $5^d = 3$. To understand this it is helpful to observe that on each of E and F we can find probability measures which give mass 3^{-j} to the part of E or F (as appropriate) inside each of the intervals obtained at the jth level of the construction.

Both of these spaces are BPI. For E this is a special case of the discussion in Section 2.3. Although the construction of F is not as homogeneous as the construction of E, it is easy to see that it is still BPI, because every piece of F inside one of the intervals obtained in the construction is an exact replica of F itself.

There is an obvious mapping $\psi : E \to F$ obtained by "sliding". A point in E can be described exactly by a sequence $\{x_j\}$ which takes values in $\{1, 3, 5\}$, this sequence describing the history of the intervals in the construction that contains the given point. We can take this sequence and modify it to get a sequence $\{y_j\}$ which takes values in $\{1, 4, 5\}$, simply by replacing each occurrence of 3 with 4 and leaving the 1's and 5's alone. This new sequence $\{y_j\}$ determines a point in F in the obvious way, by specifying again a history of intervals in the construction.

The mapping is not one-to-one. Indeed, two sequences $\{y_j\}$ give rise to the same point in F if they agree up to a certain point, and then one has a 4 followed by all 5's, and the other has a 5 followed by all 1's at that point. Thus the failure of injectivity occurs at all scales and locations. The mapping is surjective, however.

It is not hard to see that our mapping ψ is actually Lipschitz. Given distinct points p, q in E, there is an interval of length 5^{-j} in the construction which contains them and which has the property that none of its children contains both p and q. The construction ensures then that

$$5^{-j-1} \leq |p - q| \leq 5^{-j}. \tag{11.11}$$

The images of these two points under ψ lies in an interval of length 5^{-j}, and therefore the mapping is 5-Lipschitz. In particular, E looks down on F.

One can check that ψ is not bilipschitz on any set of positive measure, using the proliferation of double points. Indeed, one can even show that any set on which ψ is bilipschitz must be porous (Definition 5.7).

There are other nice properties of ψ, but we shall not discuss them until later, as part of a general discussion of mappings and their structure.

Problem 11.16 *Does F look down on E? Are they BPI equivalent?*

The construction makes it easy for one to be pessimistic and say "no" to both questions, but we do not have a proof, and we seem to be missing basic technology for resolving the matter. This is a simple example that one ought to be able to understand. We can understand this mapping ψ fairly well, the problem is to decide whether there are other mappings that might respect the geometry more efficiently.

11.7 Iterating patterns with cubes

One can make general versions of some of the earlier constructions using the procedure described in Section 2.5. Let Q denote the closed unit cube in $\mathbf{R}^n$. Fix an integer $M > 1$ and subdivide Q into M^n closed cubes of size M^{-1} in the obvious manner. Denoting by $\mathcal{S}$ the resulting collection of cubes, we get a "rule" for generating a BPI set by specifying a subset $\mathcal{R}$ of $\mathcal{S}$ with at least two elements. Assume for concreteness that we have chosen a rule $\mathcal{R}$ with k elements, and let A denote the compact BPI set in Q obtained by iterating $\mathcal{R}$.

Let us think of $\mathcal{R}$ also as an abstract finite set, we can even give it a different name when we wish to consider it in this manner, we can call it F. Let F^∞ denote the "symbolic" Cantor set as described in Section 2.3. Let us equip F^∞ with the metric given by (2.4) with $a = M^{-1}$. It is easy to build a Lipschitz mapping from F^∞ onto A, just by respecting the obvious coding. These spaces are Ahlfors regular with the same dimension, and indeed a simple way to equip A with a regular measure is to push forward the obvious product measure from F^∞. In particular we have that F^∞ looks down on A. The mapping described above is not bilipschitz in general, however, because cubes in $\mathcal{R}$ are permitted to touch each other.

Let Q' denote the unit cube in $\mathbf{R}^p$ in some dimension $p \leq n$, and let us subdivide it into cubes of size M^{-1} also. This gives a collection $\mathcal{S}'$ of cubes in Q', and subsets of $\mathcal{S}'$ define rules for generating BPI sets contained in Q'. The natural projection of $\mathbf{R}^n$ onto $\mathbf{R}^p$ induces a mapping from rules in $\mathcal{S}$ to rules in $\mathcal{S}'$. If A is generated from $\mathcal{R}$ and A' is generated from the rule $\mathcal{R}'$ in $\mathcal{S}'$ obtained by projection, then the same projection maps A onto A'.

It may be that the dimension of A' is smaller than the dimension of A, because $\mathcal{R}'$ may have fewer elements than $\mathcal{R}$. If the number of elements is the same, then the dimensions are the same, and A looks down on A'.

One can make simple examples with $n = 3$, $p = 2$, and $M = 2$.

Another possibility is to take a cube in a rule $\mathcal{R}$ which does not touch any of the others and translate it to a different position, possibly touching the others. We saw an example of this in Section 11.6. In this case the set resulting from the first rule looks down on the set resulting from the second, because the obvious Lipschitz mapping obtained from this change to the rule is surjective.

Of course there are other constructions for making one space look down on another. For instance $(\mathbf{R}, |x-y|^{\frac{1}{2}})$ looks down on $\mathbf{R}^2$ with the standard metric. This comes from a well-known version of Peano curves. The constructions mentioned above have the advantage of very simple description, but one can make more interesting examples.

In any case the question remains of knowing the extent to which the combinatorics are determined by the geometry, as in Problem 11.16.

See Section 15.5 for a general construction which captures some aspects of these examples.

11.8 An observation about snowflakes

Let $p, q, r, s \in (0, 1]$ be four numbers which satisfy $p \leq q$, $r \leq s$, and

$$\frac{1}{p} + \frac{1}{q} = \frac{1}{r} + \frac{1}{s}. \tag{11.12}$$

Consider the metric spaces

$$(M, d_M(\cdot,\cdot)) = (\mathbf{R}^2, |x_1 - y_1|^p + |x_2 - y_2|^q), \tag{11.13}$$
$$(N, d_N(\cdot,\cdot)) = (\mathbf{R}^2, |x_1 - y_1|^r + |x_2 - y_2|^s). \tag{11.14}$$

These are products of snowflake curves. These spaces are Ahlfors regular with dimension equal to the common value in (11.12). They are also BPI. Let us check that neither looks down on the other unless $p = r$ and $q = s$.

So suppose that it is not true that $p = r$ and $q = s$. As soon as one of the equalities fails the other does too, because of (11.12). Without loss in generality we may assume that $q < s$, so that $p > r$.

Suppose that M looks down on N. By definition this means that there is a Lipschitz mapping from a subset of M into N whose image has positive measure.

By Lemma 11.7 we may assume that the mapping is defined everywhere, because weak tangents of M and N are all isometrically equivalent to the original spaces.

If we have a Lipschitz mapping $F : M \to N$, then we can compose it with a projection to get a Lipschitz mapping $f : M \to (\mathbf{R}, |x-y|^r)$. If we restrict this mapping to a vertical or horizontal line then we get a mapping $h : (\mathbf{R}, |x-y|^t) \to (\mathbf{R}, |x-y|^r)$, where $t = p$ or q. The main point is that $t > r$ for either choice. This implies that h is constant. (For instance, the derivative of h in the usual sense must vanish everywhere, because of the Lipschitz condition with respect to the lopsided pair of metrics.) Thus the restriction of f to an arbitrary horizontal or vertical line is constant, and we conclude that f is itself constant. Therefore the image of F is contained in a line, and in particular cannot have positive measure. This proves that M does not look down on N.

Suppose now that N looks down on M. Given a Lipschitz mapping $\phi : N \to M$, fix $x_1 \in \mathbf{R}$, and consider the Lipschitz mapping $\psi : (\mathbf{R}, |u-v|^q) \to N$ defined by $\psi(u) = \phi(x_1, u)$. We have that $q > s \geq r$ in this case, and again the ordinary derivative of ψ vanishes everywhere, and ψ is constant. Thus $\phi(x)$ depends on x_1 alone. This implies that the image of ϕ has measure zero in N, and in fact that its Hausdorff dimension is too small. Thus N does not look down on M either.

It seems reasonable to ask whether products of snowflakes can be minimal for looking down. That is, whether it can happen for such a product space P that any BPI space M on which P looks down must be BPI equivalent to P. Not for all choices of parameters, in some cases they should be able to look down on Euclidean spaces of larger topological dimension. It is not clear exactly what is the right mixture.

Problem 11.17 *Given $s_1, s_2, \ldots, s_n \in (0, 1]$, consider the metric space*

$$\left(\mathbf{R}^n, \sum_{i=1}^{n} |x_i - y_i|^{s_i}\right). \tag{11.15}$$

Under what conditions is this BPI space minimal for looking down? Is this true when $\sum_i s_i^{-1} < n+1$, for instance?

11.9 Looking down between Cantor sets

Let $k, l \geq 2$ be integers. We basically want to work with the symbolic Cantor sets associated to sets with k and l symbols, as in Section 2.3, but it will be convenient to do things slightly differently now. Namely, we shall work with unbounded versions of the Cantor sets and doubly-infinite sequences.

Thus we let M denote the set of all mappings $x : \mathbf{Z} \to \{1, \ldots, k\}$ such that $x(j) = 1$ when $-j$ is sufficiently large, and we let N denote the analogous collection of functions taking values in $\{1, \ldots, l\}$. We define metrics for M and N as follows. Given two functions x and y on the integers, we take $L(x, y)$ to be the largest integer m such that $x(j) = y(j)$ when $j \leq m$, with $L(x, y) = \infty$ when $x = y$. We define the metrics for M and N by

$$d_M(x,y) = k^{-L(x,y)}, \qquad d_N(x,y) = l^{-L(x,y)}. \tag{11.16}$$

With these choices M and N are complete metric spaces, and they are even regular of dimension 1, as we shall check in a moment. Note that we chose the parameters carefully to get them to both have dimension 1, but for the purposes of this section all that matters is that they have the same dimension. However, there is no loss in generality in restricting ourselves to the case where the common dimension equals 1, because any other case can be reduced to that one through the snowflake transform.

To check Ahlfors regularity is it convenient to define natural measures μ and ν directly on these spaces. Given an integer j, let M_j and N_j denote the set of $x \in M$ or N (as appropriate) such that $x(i) = 1$ when $i \leq j$. Thus each of M_j and N_j can be seen as an infinite direct product of the finite sets $\{1, \ldots, k\}$ or $\{1, \ldots, l\}$. We can define on each M_j and N_j natural probability measures μ_j and ν_j by simply taking the infinite products of the uniform distributions on these two finite sets. Indeed these define Borel measures on each M_j and N_j, viewed as compact metric spaces in their own right. These measures can each be extended stupidly to all of M and N by having them assign zero measure to the complement of M_j and N_j. To define μ and ν we should combine these measures intelligently. First we set $\mu'_j = k^{-j}\,\mu_j$ and $\nu'_j = l^{-j}\,\nu_j$. These measures enjoy the compatibility property that if $i < j$, then the restriction of μ'_i to M_j agrees with μ'_j there, and similarly for N. This permits us to combine the measures μ'_j and ν'_j to get measures μ and ν on M and N, respectively, such that the restriction of μ to each M_j gives back μ'_j, and similarly for ν and N.

Given $x \in M$ and $j \in \mathbf{Z}$, we have that

$$B_M(x, k^{-j}) = \{y \in M : y(i) = x(i) \text{ when } i \leq j+1\}, \tag{11.17}$$

$$\begin{aligned} \overline{B}_M(x, k^{-j}) &= \{y \in M : y(i) = x(i) \text{ when } i \leq j\} \\ &= B_M(x, k^{-j+1}), \end{aligned} \tag{11.18}$$

and similarly for N (but with k^{-j} replaced with l^{-j}). By construction we have that

$$\mu(B_M(x, k^{-j})) = k^{-j-1}, \qquad \nu(B_N(x, l^{-j})) = l^{-j-1}. \tag{11.19}$$

This implies that M and N are Ahlfors regular of dimension 1, as in Lemma 1.2. We also have that μ and ν are bounded from above and below by H^1 on M and N, respectively.

Proposition 11.18 *If either of M or N looks down on the other, then there are integers a and b such that $k^a = l^b$. In particular this is impossible when k and l are relatively prime. Conversely this arithmetic condition implies that M and N are bilipschitz equivalent to each other.*

This can be seen as a modest variant of [CP, FM], see also [F2]. The main difference is to allow Lipschitz mappings with nontrivial image instead of only bilipschitz mappings. This variant was also established in [Mh].

The last part about the arithmetic conditions implying bilipschitz equivalence is a straightforward exercise. The first point is that M can be "reformatted" to be the same as the space P of mappings from $\mathbf{Z}$ into $\{1, \ldots, k^a\}$. One can do the same for N and get the same space P, and one can check that this is compatible also with the metrics.

To prove the first part we may as well assume that M looks down on N. This means that there is a Lipschitz mapping $f : M \to N$ whose image has positive measure. We may take f to be defined everywhere because of Lemma 11.7, and because the weak tangents to M and N are always isometrically equivalent to the original M or N with the metric multiplied by a positive constant. This is not hard to check. The main points are that M and N admit transitive groups of isometries, just by permuting the various coordinates, and that the shift mappings can be used to change scales.

Thus we assume that $f : M \to N$ is Lipschitz, and we may as well require that $f(M_0)$ have positive measure. Indeed, M is the countable union of the M_j's, and so $f(M_j)$ must have positive measure for some j if $f(M)$ is to have positive measure. We can reduce to the case where $f(M_0)$ has positive measure using shift mappings.

The idea of the proof of the proposition is roughly the following. We are going to find a spot in the image where f behaves as though it is almost measure-preserving, except for a scale factor. We shall then try to compute the measure of a ball in the image in two different ways, directly through the measure on N and indirectly through f and the measure on M. The geometry of these spaces forces a kind of discreteness which will permit us to express these measures in terms of large powers of k and l. These quantities will necessarily be approximately the same, from which we shall be able to derive the arithmetic condition above.

In fact we shall see later in Proposition 12.16 that it is possible to reduce first to the case where f is measure-preserving, modulo a scale factor, through general means. The present argument is simpler and provides a good warm-up exercise for that result.

By definitions μ_0 agrees with μ on M_0 and vanishes off of M_0. Let σ denote the measure on N obtained by pushing μ_0 forward by f, so that

$$\sigma(A) = \mu_0(f^{-1}(A)) = \mu(f^{-1}(A) \cap M_0) \tag{11.20}$$

for all Borel sets A in N. We shall measure the behavior of f through σ.

Lemma 11.19 $\sigma(N) = 1$, *and there is a constant* $C > 0$ *so that*

$$\sigma(B_N(y, r)) \geq C^{-1} r \tag{11.21}$$

when $y \in f(M_0)$ *and* $0 < r < 1$.

We have $\sigma(N) = 1$ automatically since μ_0 is a probability measure. As for the inequality, if $y = f(x)$, $x \in M_0$, then $f(B_M(x, L^{-1} r)) \subseteq B_N(y, r)$, where L is the Lipschitz constant for f. Of course we may as well take $L \geq 1$. This ensures that $B_M(x, L^{-1} r) \subseteq M_0$, since $x \in M_0$, by definition of the metric for M. Thus

$$\begin{aligned} \sigma(B_N(y,r)) &= \mu(f^{-1}(B_N(y,r)) \cap M_0) \\ &\geq \mu(B_M(x, L^{-1} r)) \geq C^{-1} r. \end{aligned} \tag{11.22}$$

This proves the lemma.

Set $F = f(M_0)$, so that F is a compact subset of N with positive ν measure. Thus the lower bound in Lemma 11.19 occurs on a set of points of positive measure, which implies a kind of smoothness for σ. We also need to control the places where σ becomes too large.

Lemma 11.20 *There is a $\lambda > 0$ (large) and a compact subset F_0 of F such that $\nu(F_0) > 0$ and*

$$\sigma(B) \leq \lambda \nu(B) \tag{11.23}$$

for all balls B in N which intersect F_0.

To prove this we shall use the following simple fact.

Sublemma 11.21 (Nesting of balls) *Given a pair of balls in M or in N, either they are disjoint or one is contained in the other.*

This is easy to derive from (11.17).

Let $\lambda > 0$ be large, to be chosen soon. Let $\mathcal{B}$ denote the collection of all balls B in N such that (11.23) fails to hold. Let $\mathcal{B}'$ denote the collection of maximal elements of $\mathcal{B}$. Every element of $\mathcal{B}$ is contained in a maximal element, and the nesting properties of balls implies that the maximal elements are pairwise disjoint. Set

$$F_0 = F \setminus \bigcup_{B \in \mathcal{B}} B = F \setminus \bigcup_{B \in \mathcal{B}'} B. \tag{11.24}$$

Let us estimate the measure of $\bigcup_{B \in \mathcal{B}'} B$. Since the elements of $\mathcal{B}'$ are pairwise disjoint and satisfy the negation of (11.23), we obtain

$$\begin{aligned} \nu\Big(\bigcup_{B \in \mathcal{B}'} B\Big) = \sum_{B \in \mathcal{B}'} \nu(B) &< \lambda^{-1} \sum_{B \in \mathcal{B}'} \sigma(B) \\ &< \lambda^{-1} \sigma(N) = \lambda^{-1}, \end{aligned} \tag{11.25}$$

by Lemma 11.19. Thus $\nu(F_0) > 0$ as soon as $\lambda > \nu(F)^{-1}$. This proves the lemma.

Let λ be fixed now, in such a way that the conclusions of the lemma hold.

We now want to modify σ off of F_0 in the Calderón–Zygmund way, to get a measure that behaves like σ near F_0 but which also behaves well everywhere. We define this new measure σ_0 by

$$\sigma_0(A) = \sigma(A \setminus \bigcup_{B \in \mathcal{B}} B) + \sum_{B \in \mathcal{B}'} \frac{\sigma(B)}{\nu(B)} \nu(A \cap B). \tag{11.26}$$

A key point now is that

$$\frac{\sigma(B)}{\nu(B)} \le l\,\lambda, \tag{11.27}$$

where l is as in the definition of N. This follows from the maximality of B; the parent of B should satisfy (11.23), and this bound follows from that, because the parent of B has measure equal to l times the measure of B (with respect to ν), as in (11.19).

Let us check that

$$\sigma_0(B) \le l\,\lambda\,\nu(B) \tag{11.28}$$

for all balls B in N. If $B \in \mathcal{B}$, then B is contained inside a unique element of $\mathcal{B}'$, and this estimate follows from the preceding one and the definition of σ_0. If B is not an element of $\mathcal{B}$, then it can be realized as the disjoint union of the elements of $\mathcal{B}'$ contained inside it and the part disjoint from $\bigcup_{B\in\mathcal{B}} B$, and in this case we get that

$$\sigma_0(B) = \sigma(B), \tag{11.29}$$

by definition of σ_0. Since B does not lie in $\mathcal{B}$ it satisfies (11.23), which implies the required estimate.

In fact we have that

$$\sigma_0(A) \le l\,\lambda\,\nu(A) \tag{11.30}$$

for all Borel sets A in N. It suffices to check this for open sets, since ν is Borel regular. (One can derive the latter from its construction or using the general Theorem 2.18 in [R].) An open subset of N can be decomposed into the disjoint union of balls contained inside of it, permitting us to reduce the present estimate to the preceding one.

We conclude in particular that σ_0 is absolutely continuous with respect to ν, and so can be represented as

$$\sigma_0 = \phi\,\nu \tag{11.31}$$

for some nonnegative function ϕ on N. In fact the preceding estimate implies that ϕ is bounded.

Applying Lebesgue's theorem to ϕ we obtain that

$$\lim_{r\to 0} \frac{1}{\nu(B_N(y,r))} \int_{B_N(y,r)} \phi(z)\,d\nu(z) = \phi(y) \tag{11.32}$$

for ν-almost all $y \in N$. For $y \in F_0$ we have that $B_N(y,r)$ is not an element of $\mathcal{B}$ (by definition of F_0), and hence

$$\sigma_0(B_N(y,r)) = \sigma(B_N(y,r)). \tag{11.33}$$

This permits us to rewrite our density formula as

$$\lim_{r\to 0} \frac{\sigma(B_N(y,r))}{\nu(B_N(y,r))} = \phi(y) \tag{11.34}$$

for ν-almost all $y \in F_0$. We can also use Lemma 11.19 to get a lower bound for these density ratios when $y \in F_0$, from which we conclude that $\phi(y) \geq C^{-1}$ ν-almost everywhere on F_0.

Let us now fix, once and for all, a point $y \in F_0$ which satisfies (11.34) and $\phi(y) > 0$. We can do these things since F_0 has positive measure.

Let $r > 0$ be very small. Choose integers p and q such that

$$l^{q-1} < r \leq l^q \quad \text{and} \quad L\,k^p \leq l^q < L\,k^{p+1}, \tag{11.35}$$

where k, l are as in the definition of M and N and L is the Lipschitz constant for f. Consider the ball $B = B_N(y, r)$. This is the same as $B_N(y, l^q)$, by the definition of the metric for N. Since f is Lipschitz $f^{-1}(B)$ is an open set, and in fact it can be realized as the disjoint union of balls of radius k^p in M. Indeed, if $x \in M$ has the property that $f(x) \in B$, then $f(B_M(x, k^p)) \subseteq B$. This is because $f(B_M(x, k^p)) \subseteq B_N(f(x), L\,k^p) \subseteq B_N(f(x), l^q)$, by the Lipschitz condition and our choice of p and q, while $B_N(f(x), l^q) \subseteq B = B_N(y, l^q)$ follows from $f(x) \in B$ and the definition of the metric on N (namely the fact that it is an ultrametric).

Thus $f^{-1}(B)$ is the disjoint union of balls in M of radius k^p. If r is small, so that p is negative, we have also that $f^{-1}(B) \cap M_0$ is the disjoint union of a similar collection of balls. (Indeed, the distance between points in M_0 and elements of its complement is always ≥ 1.) Thus

$$\sigma_0(B) = \sigma(B) = \mu(f^{-1}(B) \cap M_0) = I \cdot k^{p-1} \tag{11.36}$$

for some nonnegative integer I, because of (11.19). On the other hand we have that

$$\nu(B) = l^{q-1}, \tag{11.37}$$

again by (11.19).

We are free to choose r here, so long as it is small and positive. Let us begin by observing that

$$\left|\frac{\sigma_0(B_N(y,r))}{\nu(B_N(y,r))} - \phi(y)\right| \leq \frac{\phi(y)}{2} \tag{11.38}$$

when r is small enough, because of (11.33), (11.34), and the fact that $\phi(y) > 0$.

Let us rewrite this inequality as

$$\frac{\phi(y)}{2}\,\nu(B_N(y,r)) \leq \sigma_0(B_N(y,r)) \leq \frac{3\,\phi(y)}{2}\,\nu(B_N(y,r)). \tag{11.39}$$

Coming back to our previous story we get that

$$\frac{\phi(y)}{2}\,l^{q-1} \leq I \cdot k^{p-1} \leq \frac{3\,\phi(y)}{2}\,l^{q-1}. \tag{11.40}$$

On the other hand (11.35) gives $l^q < L\,k^{p+1}$, and so we conclude that

$$0 < I \leq \frac{3\,\phi(y)}{2}\,l^{-1}\,L\,k^2. \tag{11.41}$$

Keep in mind that k and L do not depend on r but I does. Thus I can be bounded independently of r. Since I is an integer this means that only finitely many different I's can arise in this manner.

In fact (11.34) implies that

$$\frac{\sigma_0(B_N(y,r))}{\nu(B_N(y,r))} - \phi(y) \tag{11.42}$$

tends to 0 as $r \to 0$. This means that

$$\frac{I \cdot k^{p-1}}{l^{q-1}} - \phi(y) \tag{11.43}$$

can be made as small as we like by taking r small enough, where I, p, and q depend on r. Let us rewrite this as

$$\frac{I \cdot k^{p-1}}{l^{q-1}} = \phi(y)\,(1+\epsilon(r)), \tag{11.44}$$

where $\lim_{r\to 0}\epsilon(r) = 0$. Let us make two choices of r, call them r and r', and let I', p', and q' be the parameters that correspond to r'. We can choose r and r' in such a way that they are as small as we wish, $|p-p'|$ and $|q-q'|$ are positive and bounded, and $I = I'$. This follows from (11.35) and the fact that there are only finitely many possibilities for I. In other words, we choose r and r' so that their ratio is bounded and bounded away from 0 and also so that $I = I'$. These bounds do not depend on the size of r. Applying (11.44) to both r and r' we get that

$$\frac{I \cdot k^{p-1}}{l^{q-1}}(1+\epsilon(r))^{-1} = \frac{I \cdot k^{p'-1}}{l^{q'-1}}(1+\epsilon(r'))^{-1}. \tag{11.45}$$

We can divide through to get

$$k^{p-p'} = l^{q-q'}\,(1+\delta), \tag{11.46}$$

where δ can be made as small as we wish by choosing r and r' sufficiently small.

While p, p', q, q' do depend on r, we chose them so that $p-p'$ and $q-q'$ remain bounded independently of r. If r is small enough, depending on this bound and on k and l, then we must have

$$k^{p-p'} = l^{q-q'}. \tag{11.47}$$

We also chose p, p', q, q' so that $p-p'$ and $q-q'$ are nonzero. This gives the desired arithmetic condition, and completes the proof of Proposition 11.18.

With regard to this proof one cannot avoid saying the word *entropy*. It is not clear how to formulate this elegantly as a general principle. The proof shows that we can uncover some special structure from the geometry, but it is not clear how to formulate this in a nice way.

Although Proposition 11.18 resolves the question of when the simplest self-similar Cantor sets look down on each other, there are other Cantor-BPI sets for which this remains unclear. For instance, let $(M, d_M(\cdot,\cdot))$ be as above, and let s and t be two positive numbers. Consider the space

$$(M \times M, \max\{d_M(x_1, y_1)^s, d_M(x_2, y_2)^t\}). \tag{11.48}$$

This is a BPI metric space for all choices of s and t. It is not immediately obvious when these spaces look down on each other. Think about the values of this distance function, for instance. The nonzero values are integer powers of k^s and k^t. This is more complicated when s and t are not rational multiples of each other. The kind of argument used before does not apply here, these various spaces (starting from a fixed M) are more similar measure-theoretically, even if there may be some significant oscillations.

11.10 Remarks

Let us think for a moment about Cantor sets and sets constructed from rules of cubes as in Sections 2.5 and 11.7.

Suppose that A_1 and A_2 have been constructed in the manner of Sections 2.5 and 11.7, with respect to some rules and some choices of n, M, and k which need not be the same. What can we say if we know that A_1 and A_2 are bilipschitz equivalent, or if one looks down on the other?

The point now is to assume that A_1 and A_2 are constructed through different values of k, say k_1 and k_2. Let us assume that they are Ahlfors regular of the same dimension. Under what conditions does an assumption like bilipschitz equivalence or looking down imply some restrictions on k_1 and k_2?

If the rules that were used to produce A_1 and A_2 consist of cubes that do not touch each other, then A_1 and A_2 are both bilipschitz equivalent to symbolic Cantor sets of the form F^∞, as in Section 2.3. For this bilipschitz equivalence one uses metrics of the form $d_a(\cdot,\cdot)$ as in (2.4), where a should be chosen correctly. In this case Proposition 11.18 provides strong restrictions on k_1 and k_2.

On the other hand, A_1 and A_2 could be standard line segments obtained with arbitrary choices of k_1 and k_2. One simply takes a rule which lines up a bunch of cubes in a row and does nothing else.

One could make A_1 and A_2 so that A_1 is a Cantor set and A_2 is a line segment, in such a way that A_1 looks down on A_2 no matter the values of k_1 and k_2.

What happens in between? When one uses rules in which some cubes do touch but we do not simply get something Euclidean? To what extent can one "see" the values of k associated to the Sierpinski carpet, gasket, or fractal tree?

This question is related to the story of "blocking sets" in Section 13.5 below.

We have concentrated in this chapter on BPI sets, but much the same questions are natural for more general sets, concerning the existence of nondegenerate mappings between them. We shall discuss this further later.

12

REGULAR MAPPINGS

In order to relate the geometry of spaces we want to look at mappings more general than bilipschitz. We can look at Lipschitz mappings, but they can be very degenerate. Regular mappings, defined below, provide a class of nondegenerate Lipschitz mappings for which we have some tools to establish their existence.

12.1 The definition and basic facts

Definition 12.1 (Regular mappings) *Let M and N be metric spaces. A mapping $f : M \to N$ is said to be* regular *if it is Lipschitz and if there is a constant $C > 0$ so that for every ball B_N in N we can cover $f^{-1}(B_N)$ with at most C balls in M with C times the radius of B_N.*

Normally we shall work with spaces that are doubling, so that we can simply ask that $f^{-1}(B_N)$ be covered by a bounded number of balls in M with the same radius as B_N.

The concept of regular mappings arose in [D2], in a slightly different formulation that is equivalent to the present one in when the domain is Ahlfors regular. (See Lemma 12.6 below.)

Bilipschitz mappings are automatically regular but the converse is not true. Roughly speaking, bilipschitz mappings are to injectivity as regular mappings are to bounded multiplicity. A simple example is provided by $f : \mathbf{R} \to \mathbf{R}$ given by $f(x) = |x|$, which satisfies the condition above with $C = 2$. The mappings ϕ and ψ defined in Sections 11.3 and 11.6 are regular and far from being bilipschitz.

The next lemmas illustrate the "nondegeneracy" of regular mappings.

Lemma 12.2 *Suppose that $f : M \to N$ is regular and surjective and that M is doubling. Then N is also doubling.*

This is an easy exercise.

Lemma 12.3 *If $f : M \to N$ is regular, then*

$$C^{-1} H^d(E) \leq H^d(f(E)) \leq C H^d(E) \tag{12.1}$$

for all subsets E of M, where C depends on d and the regularity constant for f but not on anything else.

This follows from the definitions. The upper bound is true for any Lipschitz mapping, but the lower bound uses regularity.

Lemma 12.4 *Suppose that $(N_1, d_1(\cdot,\cdot))$ and $(N_2, d_2(\cdot,\cdot))$ are metric spaces, and that $f : N_1 \to N_2$ is regular. If G is a connected subset of N_1, then*

$$\operatorname{diam}_{N_2} f(G) \geq C^{-1} \operatorname{diam}_{N_1} G, \tag{12.2}$$

where C depends only on the regularity constant of f.

In general regular mappings can decrease the diameter of a set dramatically, by sending a pair of distinct points to the same point for instance. However this can happen only when the set in question is made of much smaller islands, because of the definition of regularity. The lemma is easy to verify, and we omit the details.

Lemma 12.5 *Suppose that $f : M \to N$ is regular and surjective and that M is Ahlfors regular of dimension d. Then N is Ahlfors regular with dimension d.*

Of course if f is not surjective then the conclusion is that $f(M)$ is Ahlfors regular.

This is also pretty easy. One can check that closed and bounded subsets of N are compact, using the corresponding fact for M. Thus N is complete. The bound on the Hausdorff measure of balls follows from Lemma 12.3 and the simple fact that the inverse image of a ball under a surjective regular mapping contains a ball of proportional radius. (Actually we also need to observe here that M and N have comparable diameters, which is easy to establish using the regularity of M.)

(In fact, regular mappings do not shrink the diameters of uniformly perfect sets too much, as one can check.)

Lemma 12.6 *Let $f : M \to N$ be a Lipschitz mapping between metric spaces, and assume that M is Ahlfors regular of dimension d. Then f is regular if and only if there is a constant $C > 0$ so that*

$$H^d(f^{-1}(B_N(x, R))) \leq CR^d \tag{12.3}$$

for all $x \in N$ and $R > 0$.

The fact that regularity implies (12.3) is an easy consequence of the definitions and the Ahlfors regularity of M. Let us also observe that we can get the same bound when we know that N is Ahlfors regular but we do not know the same for M. That is, we use Lemma 5.1 to say that we can cover $B_N(x, R)$ by $\leq C\,(R/r)^d$ balls in N of radius r, where $0 < r < R$ is arbitrary. Using this and the regularity of f one obtains that $f^{-1}(B_N(x, R))$ can be covered with $\leq C\,(R/r)^d$ balls in M of radius r. The bound on Hausdorff measure follows easily.

Conversely, suppose that we have (12.3), and we want to show that f is regular. Fix $x \in N$ and $R > 0$. We may as well assume that $R < \operatorname{diam} M$. Let A be any subset of $f^{-1}(B_N(x, R))$ such that distinct points in A are at distance at least R from each other. Then the balls $B_M(a, R/2)$, $a \in A$, are disjoint. If f

is L-Lipschitz, then $f(B_M(a, R/2)) \subseteq B_N(x, (1 + L/2)R)$ for each $a \in A$. Using Ahlfors regularity on M we get that

$$\sum_{a \in A} R^d \leq C \sum_{a \in A} H^d(B_M(a, R/2)) \tag{12.4}$$
$$\leq CH^d(f^{-1}(B_N(x, (1 + L/2)R))) \leq C(L)R^d.$$

Thus A has a bounded number of elements. If we take A so that it has a maximal number of elements, then we must have

$$f^{-1}(B_N(x, R)) \subseteq \bigcup_{a \in A} B_M(a, R). \tag{12.5}$$

This proves that f is regular, and the lemma follows.

Lemma 12.7 (Limits of regular mappings) *Suppose that we are given a sequence of mapping packages which converges to another mapping package, as in Definition 8.18. Suppose also that the mappings in the sequence are all regular, with uniformly bounded regularity constants. Then the limiting mapping is regular too. Similarly, any weak tangent to a regular mapping (between metric spaces which are doubling and complete) is regular.*

We already know that the Lipschitz condition behaves properly in the limit (Proposition 8.20). It is not hard to derive from the definition of convergence of mappings that the regularity property persists in the limit. For the last part, about weak tangents, one need only observe that the regularity constants are preserved under the rescalings that one makes in the definition of weak tangents (because the rescalings on domain and range are the same).

For the question of limits, it is pleasant to think of regular mappings as Lipschitz mappings with the following extra property: given a ball B in the image with radius r, one cannot find more than C elements of $f^{-1}(B)$ at mutual distance $> C\,r$. This is just a reformulation of the definition.

Regular mappings are pretty nice. In this chapter we provide some technology for finding them. In Chapter 13 we shall discuss further the geometric consequences of the existence of regular mappings.

12.2 Regular mappings as weak tangents

Proposition 12.8 *Let M be an Ahlfors regular metric space of dimension d, let N be a metric space which is complete and doubling, and let Z be a compact subset of M. Suppose that $f : Z \to N$ is Lipschitz and*

$$H^d(f(Z)) > 0. \tag{12.6}$$

Then there exist weak tangents $T \in WT(M)$, $S \in WT(Z)$, and $W \in WT(N)$ such that S and T are isometrically equivalent (through a mapping which preserves basepoints) and a regular *mapping $g : S \to W$ with $g \in WT(f)$.*

The basic idea behind the proof of this is to blow up f at points where f distributes mass in a fairly nice manner. The precise argument is a little technical but not horribly difficult.

Let M, N, Z, and f be as above. Let μ denote the measure on N which is the push-forward by f of H^d on M restricted to Z, i.e.,

$$\mu(E) = H^d(f^{-1}(E) \cap Z). \tag{12.7}$$

Thus $\mu(N) < \infty$. We want to get rid of the points where μ becomes too concentrated, but first we show that they are not too numerous.

Set

$$L(\mu)(x) = \limsup_{r \to 0} r^{-d} \mu(B_N(x, r)). \tag{12.8}$$

This is a limiting version of the usual maximal function.

Lemma 12.9 *We have that*

$$H^d(\{x \in N : L(\mu)(x) > \lambda\}) \leq C \, \frac{\mu(N)}{\lambda} \tag{12.9}$$

for some $C > 0$ and all $\lambda > 0$.

When N is Ahlfors regular of dimension d we can get a similar bound for the *maximal* function. By using instead the upper limit we get a bound no matter the structure of N.

Indeed, let $\lambda > 0$ be given, set $E_\lambda = \{x \in N : L(\mu)(x) > \lambda\}$, and let $\delta > 0$ be given as well. To control $H^d(E_\lambda)$ we have to find a good covering of E_λ by sets of diameter $\leq \delta$, etc. For each $x \in E_\lambda$ we can find $0 < r < \delta$ such that $\mu(B_N(x, r)) > \lambda \, r^d$. As in Section 5.2 we can find a sequence of closed balls $\{B_i\}$ which are pairwise disjoint such that radius $B_i < \delta$ and $\mu(B_i) > \lambda \, (\text{radius}\, B_i)^d$ for all i, and either $E_\lambda \subset \bigcup_i 5B_i$ or $\limsup_{i \to \infty} \text{radius}\, B_i > 0$. From the disjointness and the lower bound on the μ-measures of the B_i's we have that

$$\sum_i (\text{radius}\, B_i)^d \leq \lambda^{-1} \mu\Big(\bigcup B_i\Big) \leq \lambda^{-1} \mu(N) < \infty. \tag{12.10}$$

In particular the radii do tend to zero, and we have a covering of E_λ by the balls $5B_i$, which are sets of diameter $< 10\,\delta$. This means that

$$\begin{aligned} H^d_{10\,\delta}(E_\lambda) \leq \sum_i (\text{diam}\, 5B_i)^d &\leq \sum_i (10 \text{ radius } B_i)^d \\ &\leq 10^d \lambda^{-1} \mu(N). \end{aligned} \tag{12.11}$$

Since this holds for all $\delta > 0$ we get the bound asserted in the lemma.

Lemma 12.9 is not quite good enough for our purposes, we need to have a lot of points where the concentration of μ is bounded uniformly at small scales. The next lemma will give us that. Given $\rho > 0$, set

$$L_\rho(\mu)(x) = \sup_{0<r<\rho} \frac{\mu(B_N(x,r))}{r^d}. \tag{12.12}$$

For measurability purposes it is helpful to notice that the supremum can be restricted to rational values of r.

Lemma 12.10 *We have*

$$\lim_{\rho\to 0} H^d(\{x \in f(Z) : L_\rho(\mu)(x) > \lambda\}) \leq C\,\frac{\mu(N)}{\lambda} \tag{12.13}$$

for some $C > 0$ and all $\lambda > 0$.

Indeed, $\{x \in f(Z) : L_\rho(\mu)(x) > \lambda\}$ is a family of Borel sets of *finite* measure (since in the image of f) which decreases, as $\rho \to 0$, to a subset of $\{x \in f(Z) : L(\mu)(x) \geq \lambda\}$. Thus Lemma 12.10 follows from Lemma 12.9.

Set $E_{\rho,\lambda} = \{x \in f(Z) : L_\rho(\mu)(x) > \lambda\}$. Combining Lemma 12.10 with (12.6) we get that

$$H^d(f(Z)\backslash E_{\rho,\lambda}) > 0 \tag{12.14}$$

if we choose $\lambda > 0$ large enough and then $\rho > 0$ small enough. Fix λ and ρ with these properties, and set $G = Z\backslash f^{-1}(E_{\rho,\lambda})$. Thus $H^d(f(G)) > 0$, and $H^d(G) > 0$ in particular. Notice that G is compact, because $E_{\rho,\lambda}$ is relatively open in $f(Z)$.

Fix a point of density $x_0 \in G$. For $r > 0$ small consider the pointed metric spaces $M_{x_0,t}$, $Z_{x_0,t}$, and $N_{f(x_0),t}$, as in Section 9.1. We can view f as a mapping from $Z_{x_0,t}$ to $N_{f(x_0),t}$, and as such it preserves the basepoints and is uniformly Lipschitz (i.e., the Lipschitz constant is bounded independently of t).

Claim 12.11 *There is a sequence $\{t_j\}$ of positive numbers which tend to 0 such that M_{x_0,t_j} converges to a weak tangent $T \in WT(M)$, Z_{x_0,t_j} converges to a weak tangent $S \in WT(Z)$, $N_{f(x_0),t_j}$ converges to a weak tangent $W \in WT(N)$, S is isometrically equivalent to T (with the basepoints preserved), and the mapping packages $(Z_{x_0,t_j}, N_{f(x_0),t_j}, f)$ converge to a mapping package (S, W, g).*

This follows from our usual existence theorems. We start by using Lemmas 9.12 and 9.13 to say that we can get tangents S and T as above (isometrically equivalent in particular). We use Lemma 8.22 for the existence of the limit of mapping packages. Of course we pass to subsequences whenever necessary. For the existence of the limit of mapping packages note that the uniform Lipschitz bounds on our mappings imply the required equicontinuity and uniform boundedness conditions.

We want to show that this mapping $g : S \to W$ is regular. This will do the job, since S and T are isometrically equivalent.

Fix a ball $B = B_W(p,r)$ in W. We want to show that $g^{-1}(B)$ can be covered by a bounded number of balls of radius r in S.

Let $\phi_j : W \to N_{f(x_0),t_j}$ and $\psi_j : N_{f(x_0),t_j} \to W$ be as in Lemma 8.11. These are basepoint-preserving mappings which become approximately isometric as $j \to \infty$.

Let B_j be the ball in $N_{f(x_0),t_j}$ with center $p_j = \phi_j(p)$ and radius r with respect to the metric of $N_{f(x_0),t_j}$, which means radius $r\,t_j$ with respect to the metric of N. We think of this ball as being an approximation to B. We want to show that $f^{-1}(B_j)$ can be covered by a bounded number of balls of the correct radius when j is large enough. "Correct radius" means $r\,t_j$ with respect to our original metric on M. Actually we do not care about all of $f^{-1}(B_j)$, just the part that is not too far from x_0 compared to t_j. (There might be some bad parts that leak off to infinity in M_{x_0,t_j} as $j \to \infty$.) In order to find this covering we first control the measures of these sets. Before doing that let us record some useful consequences of the fact that x_0 is a point of density of G (and hence Z).

Lemma 12.12 *There is a sequence $\{k_j\}$ of positive integers with $k_j \to \infty$ as $j \to \infty$ such that*

$$\lim_{j\to\infty} t_j^{-d}\, H^d(B_M(x_0,(2\,k_j+1)\,t_j)\backslash G) = 0 \tag{12.15}$$

and

$$\lim_{j\to\infty} \sup_{y\in B_M(x_0,2\,k_j\,t_j)} t_j^{-1}\,\mathrm{dist}(y,G) = 0. \tag{12.16}$$

Indeed, since x_0 is a point of density of G, we have that

$$\lim_{s\to 0} s^{-d}\, H^d(B_M(x_0,s)\backslash G) = 0. \tag{12.17}$$

Using this it is easy to choose $\{k_j\}$ so that (12.15) holds. This implies (12.16), because of Ahlfors regularity. This proves the lemma.

Let $\{k_j\}$ be fixed as above, from now on.

Lemma 12.13 *For sufficiently large j we have that*

$$H^d(f^{-1}(2B_j)\cap B_M(x_0,2\,k_j\,t_j)) \le C\,(t_j\,r)^d, \tag{12.18}$$

where C does not depend on j or our initial choice of ball $B = B_W(p,r)$.

The Hausdorff measure in this inequality is taken with respect to our original metric on M, not any of the rescaled metrics. This corresponds to the presence of the t_j on the right-hand side. Note that $f^{-1}(2B_j)$ (or the inverse image of any subset on N) is automatically a subset of Z, the domain of f.

Set $H_j = f^{-1}(2B_j)\cap B_M(x_0,2\,k_j\,t_j)$, and remember that it lives in M. We may as well assume that $H_j \neq \varnothing$, since otherwise the lemma is trivial. Let L be the Lipschitz norm of f (with respect to the original metrics). If j is large enough, then there is a point $v_j \in G$ such that

$$\mathrm{dist}(v_j,H_j) \le t_j\,r\,L^{-1}, \tag{12.19}$$

because of (12.16). From this and the definition of H_j we get that

$$H_j \subseteq f^{-1}(B_N(f(v_j), 5\,r\,t_j)) \cap B_M(x_0, 2\,k_j\,t_j), \tag{12.20}$$

since $2B_j$ has radius $2\,r\,t_j$ with respect to the original metric on N. Because $v_j \in G$ we have that $f(v_j) \notin E_{\rho,\lambda}$, and hence

$$\mu(B_N(f(v_j), 5\,r\,t_j)) \leq \lambda\,(5\,r\,t_j)^d \tag{12.21}$$

if j is large enough. Lemma 12.13 follows now by combining (12.20) with this inequality and using the definition of μ as the push-forward by f of the restriction of H^d to Z (= the domain of f).

Lemma 12.14 *We can cover $f^{-1}(B_j) \cap B_M(x_0, k_j\,t_j)$ by at most C balls of radius $r\,t_j$ centered on the set when j is large enough. Here C does not depend on j or our initial choice of ball $B = B_W(p, r)$, and we are using the original metric on M for defining the radii.*

Let j be large, and let A be a subset of $f^{-1}(B_j) \cap B_M(x_0, k_j\,t_j)$ with the property that $d_M(x, y) \geq r\,t_j$ when $x, y \in A$, $x \neq y$, and write $\#A$ for the number of elements of A. We want to obtain a bound for $\#A$. Let L denote the Lipschitz constant for f, as before. For j sufficiently large we have that

$$H^d(B_M(x, (L+2)^{-1}\,r\,t_j) \cap Z) \geq C^{-1}\,(r\,t_j)^d \qquad \text{for all } x \in A, \tag{12.22}$$

where C depends on L but not on j or $B = B_W(p, r)$. This follows from (12.15) (and $G \subseteq Z$). Hence

$$\begin{aligned} \#A \cdot (r\,t_j)^d &\leq C \sum_{x \in A} H^d(B_M(x, (L+2)^{-1}\,r\,t_j) \cap Z) \\ &\leq C\,H^d\Big(\bigcup_{x \in A} B_M(x, (L+2)^{-1}\,r\,t_j) \cap Z\Big), \end{aligned} \tag{12.23}$$

using (12.22) in the first inequality and the disjointness of the balls in the second. On the other hand we have that

$$\bigcup_{x \in A} B_M(x, (L+2)^{-1}\,r\,t_j) \cap Z \subseteq f^{-1}(2B_j) \cap B_M(x_0, 2\,k_j\,t_j) \tag{12.24}$$

since $A \subseteq f^{-1}(B_j) \cap B_M(x_0, k_j\,t_j)$ (and $f^{-1}(B_j) \subseteq Z$ automatically, Z being the domain of f). Thus

$$\#A \cdot (r\,t_j)^d \leq C\,H^d(f^{-1}(2B_j) \cap B_M(x_0, 2\,k_j\,t_j)), \tag{12.25}$$

from which we conclude that the number of elements of A is bounded, because of Lemma 12.13. That is to say, the number of elements is bounded if j is large enough, with a bound that does not depend on j or our ball $B = B_W(p, r)$.

This bound on the number of elements of A implies that we can choose A to have a maximal number of elements, with a bound on that number. For such a maximal A we have that

$$f^{-1}(B_j) \cap B_M(x_0, k_j t_j) \subseteq \bigcup_{x \in A} B_M(x, r t_j). \tag{12.26}$$

This proves Lemma 12.14.

Lemma 12.14 can be reformulated as saying that $f^{-1}(B_j) \cap B(x_0, k_j)$ can be covered by a bounded number of balls of radius r when j is sufficiently large, where now we are using the metric for Z_{x_0,t_j} to define $B(x_0, k_j)$ and the radii of these balls. This fact and a simple convergence argument implies that $g^{-1}(B)$ is covered by a bounded number of balls of radius r. Thus g is regular, as desired. This completes the proof of Proposition 12.8.

12.3 Looking down from $\mathbf{R}^n$

In this section we give another proof of Proposition 11.10.

Let M be a BPI space of dimension n, and suppose that $\mathbf{R}^n$ looks down on M. We want to show that M is BPI equivalent to $\mathbf{R}^n$.

By assumption there is a closed subset A of $\mathbf{R}^n$ and a Lipschitz mapping $f : A \to M$ such that $f(A)$ has positive measure in M. Of course we may as well assume that A is compact. Using Proposition 12.8 we obtain the existence of a regular mapping from all of $\mathbf{R}^n$ into a weak tangent M' of M. (We can take the domain to be all of $\mathbf{R}^n$ because any weak tangent to $\mathbf{R}^n$ is isometrically equivalent to $\mathbf{R}^n$.) From Proposition 9.9 we know that M' is BPI and BPI equivalent to M, and Proposition 7.5 implies that M is BPI equivalent to $\mathbf{R}^n$ if M' is. Thus we may as well assume from the start that we have a regular mapping from $\mathbf{R}^n$ into M.

In fact we may as well assume that we have a regular mapping onto M. The reason is that the image of f will be a regular subset of M, by Lemma 12.5, and regular subsets of BPI sets are BPI sets (Proposition 6.8) which are BPI equivalent to the original set (Proposition 7.1).

Next we claim that M has topological dimension equal to n. We shall not recall the definition here but simply refer the reader to [HW], which contains all the information about topological dimension that we shall need. To prove this claim we first use the fact that M has Hausdorff dimension n, and so has topological dimension $\leq n$ (p.104 of [HW]). Suppose that M had topological dimension less than n. Then Theorem VI 7 on p.91 would imply that there is a point $y \in M$ such that $f^{-1}(y)$ has positive topological dimension. This uses the fact that $\mathbf{R}^n$ has dimension exactly n, and the fact that f is a "closed" mapping, which means that it maps closed sets to closed sets. The latter is an easy consequence of regularity. From regularity we also have that $f^{-1}(y)$ is a finite set for any $y \in M$. This is not hard to check. This contradicts the assertion that $f^{-1}(y)$ has positive topological dimension, and so we conclude that the topological dimension of M must be equal to n.

Using now Theorem VI 2 on p.77 of [HW] we obtain the existence of a continuous mapping ϕ from M into a closed n-dimensional cube in $\mathbf{R}^n$ which has a stable value u. This means that there is a $\delta > 0$ so that if ψ is any other continuous mapping from M into the same n-cube such that ϕ and ψ are at distance $\leq \delta$ in the supremum norm, then ψ also takes the value u. We can approximate ϕ in the supremum norm by functions which are locally Lipschitz, and so we may as well assume that ϕ is locally Lipschitz. ("Locally Lipschitz" because we are working on a noncompact space. On any metric space we can approximate uniformly continuous functions by Lipschitz functions uniformly. We can do this for all continuous functions on a compact space, but on a noncompact space we should allow locally Lipschitz functions if we want a uniform approximation. This technical point could be simplified by making a localization to compact spaces first, but this is not necessary.) The stability condition ensures that the image of ϕ actually includes a neighborhood of u in the n-cube.

The conclusion of these results from dimension theory is that there is a locally Lipschitz mapping ϕ from M into $\mathbf{R}^n$ whose image has positive measure. Thus we can compose with f and get a locally Lipschitz mapping $\phi \circ f$ whose image has positive measure. It is a standard fact that such a mapping is bilipschitz on a subset of positive measure. (For instance, one can use the fact that for each $\mathbf{R}^n$-valued Lipschitz mapping on $\mathbf{R}^n$ there is a C^1 mapping which agrees with it except on a set of small measure. One can force this agreement to hold on a set whose image has positive measure and then apply the inverse function theorem to the C^1 mapping to get a nontrivial bilipschitz piece for the original mapping.)

Once we know that $\phi \circ f$ is bilipschitz on a set of positive measure, we conclude that f is bilipschitz on a set of positive measure, since ϕ is locally Lipschitz. This implies that M is BPI equivalent to $\mathbf{R}^n$, as in Proposition 7.1.

This completes the proof.

This proof illustrates a nice principle, which is that something like the topological dimension should not decrease when one has a BPI space looking down on another one. (It can easily increase, as in the example of a Cantor set looking down on the real line.) One has to be a little careful in making a precise statement, because of the passage to weak tangents. We shall not try to beat out a good formulation here, the principle is clear enough from the argument above, but if one is interested in such things one should take into account "uniform" versions of topological dimension as discussed in [Se6].

12.4 Measure-preserving weak tangents

In Proposition 12.8 we saw that we could find regular mappings among the weak tangents of a mapping under a mild nondegeneracy condition. In this section we wish to establish a similar result in which the weak tangent satisfies a stronger measure-preserving property. For this purpose we need some additional terminology about weak tangents. We start with some basic assumptions that will be in force throughout this section.

Assumptions 12.15 *M and N are metric spaces, M is Ahlfors regular and N is doubling and complete, μ is a d-dimensional Ahlfors regular measure (Definition 8.27) on M whose support is all of M, ν is a d-dimensional Ahlfors regular measure on N, and $f : M \to N$ is Lipschitz.*

Let us call a *package* a collection $\mathcal{P}$ consisting of a pair of pointed metric spaces S and T which are doubling and complete together with a Lipschitz mapping $g : S \to T$ and a pair of d-dimensional Ahfors regular measures σ and τ, where σ lives on S and has support equal to S, and τ lives on T.

We shall write WT for the collection of packages which are *weak tangents* to our initial data in Assumptions 12.15. This means packages which arise as weak limits of rescalings of our initial package, in the following sense. We start with a sequence $\{p_j\}$ in M and a sequence of positive numbers $\{t_j\}$. As before we ask that $t_j \leq \min(\operatorname{diam} M, \operatorname{diam} N)$ for each j. Set $q_j = f(p_j)$, let M_{p_j,t_j} and N_{q_j,t_j} be as in Section 9.1, and set $\mu_{p_j,t_j} = t_j^{-d}\mu$, $\nu_{q_j,t_j} = t_j^{-d}\nu$, viewed as measures on M_{p_j,t_j} and N_{q_j,t_j}. We can also view f now as a Lipschitz mapping from M_{p_j,t_j} to N_{q_j,t_j}, with the same Lipschitz constant as before. A package $\mathcal{P} = (S, T, g, \sigma, \tau)$ is said to lie in WT if it arises as a limit of the sequence of packages $(M_{p_j,t_j}, N_{q_j,t_j}, f, \mu_{p_j,t_j}, \nu_{q_j,t_j})$ for some choice of $\{p_j\}$, $\{t_j\}$. (One can define "limits of sequences of packages" in the obvious way, combining our earlier definitions.)

Of course we can define this version of weak tangents for any quintuple of metric spaces, mapping, and measures as above, and we can apply the notion to the elements of WT. As usual we have that limits of elements of WT also lie in WT. This fact will play an important role in the argument that follows. Its proof is straightforward but unpleasant to read or write, and we omit the details.

This notion of weak tangent of packages is a bit convoluted when one goes back to our definitions of convergence, relying on choices of embeddings into Euclidean spaces and so forth, but in fact we may as well restrict ourselves to the case where everything lives inside of some fixed $\mathbf{R}^n$. We shall come back to this point after stating the main result of this section.

Proposition 12.16 (Measure-preserving weak tangents) *Assumptions as above. If also $\nu(f(M)) > 0$, then there exists a package $\mathcal{P}_0 = (S_0, T_0, g_0, \sigma_0, \tau_0)$ in WT and a constant $\theta > 0$ such that $\tau_0(g_0(K)) = \theta\, \sigma_0(K)$ for all compact subsets K of S. (Compactness is assumed to avoid minor issues of measurability.)*

Note that the measure-preserving condition need not imply anything like bilipschitzness, although it is a substantial nondegeneracy condition. We shall discuss the measure-preserving condition further in the next section, concentrating on the proof of this existence result in this section.

For the purposes of Proposition 12.16 we may as well assume that M and N are closed subsets of some $\mathbf{R}^n$. This is because the assumptions and conclusions of the proposition are not disturbed if we apply snowflake transforms to M and N (of the same order, so that dimensions are preserved), or if we replace M and N by bilipschitz equivalent spaces. Thus we can use Assouad's embedding theorem to put everything inside some $\mathbf{R}^n$. This reduction is not truly significant but it is convenient when one tries to sort out the details of the definition of weak tangents. By simply using subsets of Euclidean spaces we can work with limits of subsets of Euclidean spaces rather than limits of metric spaces, thereby avoiding the nuisance of the embeddings, we can take limits of measures directly on $\mathbf{R}^n$ as well, etc. This also makes it easier to check that weak tangents of weak tangents are weak tangents.

In a technical sense this definition of weak tangents, using fixed embeddings into $\mathbf{R}^n$, is not literally the same as our official definition, in which we permit the embeddings to depend on the weak tangent. We have seen before though that we do not really gain anything new by allowing other embeddings, because of the various statements of uniqueness up to isometric equivalence. Alternatively the reader can just interpret "weak tangents" in this section to mean that we are using fixed embeddings into Euclidean spaces. The main points that we shall need are the existence of limits when we pass to subsequences and the fact that weak tangents of weak tangents are weak tangents, and this is even clearer when we use fixed embeddings.

Let us now prove Proposition 12.16. We begin by choosing θ. We set

$$\theta = \sup\Big\{ \frac{\tau(g(K))}{\sigma(K)} : \mathcal{P} = (S, T, g, \sigma, \tau) \in WT, \qquad K \subseteq S \text{ is compact, and } \sigma(K) > 0\Big\}. \tag{12.27}$$

That is, we look at the supremum of the mass ratios for all weak tangents. To produce the measure-preserving weak tangent we shall look for a weak tangent which achieves the maximum.

We should observe first that θ is finite. Our definition of weak tangents ensures that g has bounded Lipschitz norm when $(S, T, g, \sigma, \tau) \in WT$, in fact that the Lipschitz norm is bounded by that of f. This implies that

$$\frac{H^d(g(K))}{H^d(K)}, \qquad K \subseteq S \text{ compact, } H^d(K) > 0, \tag{12.28}$$

remains bounded. We also know have that the Ahlfors regularity constants of σ and τ are bounded. This implies that σ and τ are each bounded above and below by constant multiples of the restriction of H^d to their respective supports, as in Lemma 1.2, and with constants that are uniformly bounded and bounded away from 0. Since we are assuming that the support of μ is all of M, we have the analogous property for σ and S, and so σ is bounded above and below by multiples of H^d on S. This permits us to conclude the finiteness of θ from the corresponding bounds for Hausdorff measure.

We should also point out that $\theta > 0$. Indeed, we can practically take our initial spaces, mapping, and measure as a package in WT, we have only to choose a basepoint in M and take its image to be our basepoint in N. In other words, we can take the sequences $\{p_j\}$ and $\{t_j\}$ in the definition of WT to be constant, so that our initial objects are included as competitors. Our nondegeneracy assumption that $\nu(f(M)) > 0$ ensures then that $\theta > 0$.

Eventually we are going to want to find a weak tangent for which the supremum in (12.27) is attained, and to do this we shall need to be able to compute limits of measures of sets. For this reason we shall want to work with *smooth sets*, as in Definition 8.36. We know from Lemma 8.37 that there are plenty of smooth sets inside of Ahlfors regular sets, namely the cubes in the sense of Section 5.5. For the moment it will even be convenient for us to restrict ourselves to cubes.

We have a small nuisance problem here, which is that cubes need not be compact, but we need to work with compact sets. We shall work with the closures of cubes instead of cubes, and call these "compact cubes". We shall abuse our terminology slightly and call two compact cubes "disjoint" if they were obtained as the closures of disjoint cubes. For the purposes of measure-theoretic computations this abuse of language is harmless, for if Q and Q' are disjoint cubes, then

$$H^d(\overline{Q}\backslash Q) = 0 \quad \text{and} \quad H^d(\overline{Q} \cap \overline{Q'}) = 0. \tag{12.29}$$

This follows from (5.9). If we are inside a package $\mathcal{P} = (S, T, g, \sigma, \tau) \in WT$, then we have

$$\begin{aligned} \sigma(\overline{Q}\backslash Q) = 0\ , \qquad & \sigma(\overline{Q} \cap \overline{Q'}) = 0, \\ \tau(g(\overline{Q}\backslash Q)) = 0\ , \qquad & \tau(g(\overline{Q} \cap \overline{Q'})) = 0, \end{aligned} \tag{12.30}$$

since both σ and τ are controlled by Hausdorff measure, and since g is Lipschitz.

Lemma 12.17 (Restriction to cubes) *If we restrict ourselves in (12.27) to subsets K of S which are compact cubes then we get the same value of θ. This remains true if we restrict ourselves to compact cubes of diameter $\leq \eta$ for any $\eta > 0$ given in advance.*

To see this we let θ' denote the analogue of θ defined using compact cubes only. Of course $\theta' \leq \theta$ by definition, the issue is to prove the reverse inequality. Fix an arbitrary package $\mathcal{P} = (S, T, g, \sigma, \tau) \in WT$ and a compact set $K \subseteq S$ with $\sigma(K) > 0$, and let $\epsilon > 0$ be given. Consider the compact cubes Q in S such that $\operatorname{diam} Q \leq \eta$ and

$$\sigma(Q \cap K) > (1 - \epsilon)\, \sigma(Q). \tag{12.31}$$

Almost every element of K is contained in a compact cube of this type, because every point of density of K is contained in such a cube. This implies that almost every point in K lies in a maximal cube with the same properties. Let $\{Q_i\}$ be a listing of the maximal compact cubes which satisfy $\operatorname{diam} Q \leq \eta$ and (12.31). Then the Q_i's are pairwise disjoint and

$$\sigma\Big(K \backslash \bigcup Q_i\Big) = 0. \tag{12.32}$$

This implies also that $\tau(g(K \backslash \bigcup Q_i)) = 0$, since σ and τ are both equivalent in size to the restriction of H^d to their supports, and since g is Lipschitz. In short we can treat K as though it were a subset of $\bigcup Q_i$ for making computations about measures. We have that

$$\begin{aligned}\tau(g(K)) &\leq \tau\Big(g\Big(\bigcup Q_i\Big)\Big) \leq \sum_i \tau(g(Q_i)) \\ &\leq \theta' \sum_i \sigma(Q_i) \leq \theta' \, (1-\epsilon)^{-1} \sum_i \sigma(Q_i \cap K) \\ &\leq \theta' \, (1-\epsilon)^{-1} \, \sigma(K).\end{aligned} \tag{12.33}$$

Since ϵ is arbitrary we get that $\tau(g(K)) \leq \theta' \, \sigma(K)$. This implies that $\theta \leq \theta'$, and Lemma 12.17 follows.

In a moment we shall look at maximizing sequences for θ, but before we do that let us make some computations. Fix a package $\mathcal{P} = (S, T, g, \sigma, \tau) \in WT$. Given a compact set $K \subseteq S$ set

$$\rho(K) = \theta - \frac{\tau(g(K))}{\sigma(K)}. \tag{12.34}$$

Thus $\rho(K)$ is always nonnegative, and we are interested in knowing when it is small.

Lemma 12.18 *Suppose that K is a compact cube in S and that F is a finite union of compact subcubes of K. Then*

$$\frac{\sigma(F)}{\sigma(K)} \, \rho(F) \leq \rho(K). \tag{12.35}$$

Indeed, given K and F as in the lemma, let E be the closure of $K \backslash F$, itself a union of compact cubes, each disjoint from the compact cubes in F. It suffices to show that

$$\sigma(F) \, \rho(F) + \sigma(E) \, \rho(E) \leq \sigma(K) \, \rho(K), \tag{12.36}$$

since the E term is nonnegative. By definition of ρ this reduces to

$$\tau(g(F)) + \tau(g(E)) \geq \tau(g(K)). \tag{12.37}$$

This inequality is automatic, and the lemma follows.

One conclusion of the lemma is that if we know that $\rho(K)$ is small, then we also know that $\rho(F)$ is small when F is a subcube of K which is not too much smaller.

Let us now choose a maximizing sequence for θ. That is, we take a sequence of packages $\mathcal{P}_l = (S_l, T_l, g_l, \sigma_l, \tau_l) \in WT$ and a sequence of compact cubes $K_l \subseteq S_l$ such that

$$\lim_{l \to \infty} \frac{\tau_l(g_l(K_l))}{\sigma_l(K_l)} = \theta. \tag{12.38}$$

We should start to be slightly more careful about the story of cubes, since we are working with sequences of spaces. The main point is that all the S's that arise from weak tangents, and all the S_l's in particular, are Ahlfors regular with bounded constants. This means that the various constants that appear in the story of cubes from Section 5.5 can be taken to be uniformly bounded.

Let us check that we may as well assume that

$$(\operatorname{diam} K_l)^d = \rho(K_l)^{-\frac{1}{3}} \qquad \text{for all } l. \tag{12.39}$$

For this we should require that $\rho(K_l)$ be small for all l, which we can certainly do, since they tend to 0 anyway. In order to get this normalization (12.39) we observe first that we could require that $\operatorname{diam} K_l \leq 1$ for all l. This comes as a by-product of Lemma 12.17, which says that we may restrict ourselves to cubes that are as small as we like. To get (12.39) we simply expand the metrics on each S_l by the necessary factor. We can do this, as in (9.1). In order to keep our package a weak tangent we need to rescale the metric on T_l in the same way, and to renormalize σ_l and τ_l accordingly, but this is all easy to do. Thus we may assume that (12.39) holds.

Notice that this rescaling is not traumatic for the notion of cubes. One must simply adjust the labelling to match the new diameters, so that the "cubes of size 2^j" (the elements of Δ_j in Section 5.5) still have size about 2^j. This is just a cosmetic issue. Of course the rescaling is harmless for the Ahlfors regularity conditions.

The rescaling does not affect (12.38) either, since the σ_l's and τ_l's will be rescaled in the same manner.

Similarly we would like to move the basepoints of the S_l's. According to (5.10), for each l we can find a point $s_l \in K_l$ such that

$$B_{S_l}(s_l, C_0^{-1} \operatorname{diam} K_l) \subseteq K_l, \tag{12.40}$$

where C_0 does not depend on l (but is controlled by the Ahlfors regularity constants). We may as well assume that the s_l's are the basepoints for the pointed metric spaces S_l. If they were not the basepoints before, we can simply consider the new pointed metric spaces where they are the basepoints. If a pointed metric space is a weak tangent for some given metric space, and if we move the basepoint of the weak tangent, then we still have a weak tangent. One simply has to check that one can adjust the basepoints in the sequence of approximations accordingly. This is not difficult. (The very definition of convergence implies that there are points in the approximating spaces to move to.) In our business with packages one has to move the basepoints for the T_l's (to $g_l(s_l)$) and to check that the whole package continues to lie in WT, but nothing mysterious happens.

With these normalizations the K_l's are now large chunks of the S_l's which are centrally located. We want to check that there are other centrally located blobs of size that we can control and for which we have good control of the mass ratios.

Lemma 12.19 *There is a constant $C > 0$ so that for all sufficiently large l and each positive integer m with $C\,2^m \leq \operatorname{diam} K_l$ we can find a compact subset*

$F(l,m)$ of S_l such that each $F(l,m)$ is a union of $\leq C$ compact subcubes of K_l and

$$B_{S_l}(s_l, 2^m) \subseteq F(l,m) \subseteq B_{S_l}(s_l, C\,2^m). \tag{12.41}$$

This follows easily from the properties of cubes. For each m one takes $F(l,m)$ to be the union of the compact subcubes of K_l of size 2^m which intersect the ball $B_{S_l}(s_l, 2^m)$. A key point is that there are only boundedly many of these subcubes. We are also using (12.40) here.

Let us check that

$$\rho_l(F(l,m)) \leq \rho_l(K_l)^{\frac{1}{2}} \tag{12.42}$$

for all sufficiently large l and all m as above. Here we have added the subscript l to ρ to reflect its dependence on the lth package. According to Lemma 12.18 it suffices to know that

$$\frac{\sigma_l(F(l,m))}{\sigma_l(K_l)} \geq \rho_l(K_l)^{\frac{1}{2}}. \tag{12.43}$$

This follows from Ahlfors regularity and our normalization (12.39) as soon as l is large enough. Thus we have (12.42).

We are now practically finished. By passing to a subsequence we may assume that our packages $\mathcal{P}_l = (S_l, T_l, g_l, \sigma_l, \tau_l) \in WT$ converge to a package $\mathcal{P} = (S, T, g, \sigma, \tau)$. This limiting package also lies in WT, because limits of weak tangents are still weak tangents.

We may also assume that the sequences of compact sets $\{F(l,m)\}_l$ converge inside the S_l's to compact subsets $F(m)$ in S. Here we are using the notion of convergence of subsets given in Definition 8.30. We are also using Lemma 8.31 to know that limits exist when we pass to a subsequence. Notice that condition (8.27) is automatic in our case, because we have normalized things so that each $F(l,m)$ contains the basepoint s_l of S_l. Lemma 8.31 says that we can pass to a subsequence to get $\{F(l,m)\}_l$ to converge for any given m, we can get convergence for all of them using a Cantor diagonalization procedure.

Lemma 12.20 $\rho(F(m)) = 0$ *for all* m.

This time $\rho(\cdot)$ refers to our limiting package $\mathcal{P} = (S, T, g, \sigma, \tau)$.

It suffices to show that

$$\rho(F(m)) \leq \liminf_{l \to \infty} \rho_l(F(l,m)), \tag{12.44}$$

since $\rho(F(m)) \geq 0$ automatically and we know that the limit on the right side vanishes, because of (12.42) and the fact that $\rho_l(K_l) \to 0$, by (12.38). Let us rewrite (12.44) as

$$\frac{\tau(g(F(m)))}{\sigma(F(m))} \geq \limsup_{l \to \infty} \frac{\tau_l(g_l(F(l,m)))}{\sigma_l(F(l,m))}. \tag{12.45}$$

To prove this it suffices to show that

$$\tau(g(F(m))) \geq \limsup_{l \to \infty} \tau_l(g_l(F(l,m))) \tag{12.46}$$

and

$$\sigma(F(m)) = \lim_{l \to \infty} \sigma_l(F(l,m)). \tag{12.47}$$

The first follows from Lemma 8.33, at least if we can verify its hypotheses. The condition (8.30) is immediate from our definitions. For Lemma 8.33 we need to know that T_l converges to T (as pointed metric spaces) and τ_l converges to τ, and we know these to be true by construction. We also need to know that $g_l(F(l,m))$ converges to $g(F(m))$ as $l \to \infty$ as subsets of T_l, T. This follows from Lemma 8.32, because of the convergence of g_l to g and $F(l,m)$ to $F(m)$. Thus we get (12.46). For (12.47) we apply Lemma 8.39. In this case the requirements of Lemma 8.39 are provided by the construction, with the small issue of the uniform bounds on the smoothness constants of the $F(l,m)$'s. This comes from the following facts: the $F(l,m)$'s are finite unions of compact cubes (Lemma 12.19), with a uniform bound on the number of cubes needed; cubes are smooth (Lemma 8.37), with uniform bounds; and smoothness behaves properly under finite unions (Lemma 8.38). Keep in mind that we have bounds on the constants which govern the behavior of cubes coming from our uniform bounds on the Ahlfors regularity constants. We maintain bounds on the smoothness constants when we take finite unions here because the smooth sets whose union is being taken have all approximately the same diameter, to within a bounded factor. (It does not matter what the diameters are, so long as they are approximately the same.) We did not state this in Lemma 12.19, but it is clear from the proof.

Thus we have (12.46) and (12.47), and the lemma follows.

We are almost finished now. Let K be any compact subset of S, and let us show that

$$\tau(g(K)) = \theta\,\sigma(K). \tag{12.48}$$

We know that

$$\tau(g(E)) \leq \theta\,\sigma(E) \tag{12.49}$$

for all compact subsets of S, because of the maximality of θ. That is, θ was defined in (12.27) by taking the supremum over all elements of WT, and we have one here. On the other hand we have that

$$\tau(g(F(m))) = \theta\,\sigma(F(m)) \tag{12.50}$$

for all m by the lemma. Also

$$B_S(s, 2^m) \subseteq F(m), \tag{12.51}$$

where s denotes the basepoint of S. This follows from (12.41), i.e., this kind of inclusion persists in the limit (because of the convergence of the metrics). Thus $K \subseteq F(m)$ if m is large enough.

Notice that

$$\tau(g(F(m)\backslash K)) \leq \theta\,\sigma(F(m)\backslash K). \tag{12.52}$$

This follows from (12.49); although $F(m)\backslash K$ need not be compact, so that we may not be able to invoke the inequality directly, it is a countable union of compact sets, which permits us to derive this from the version for compact sets. Thus we have

$$\begin{aligned}\tau(g(F(m))) &\leq \tau(g(K)) + \tau(g(F(m)\backslash K)) \\ &\leq \theta\,\sigma(K) + \theta\,\sigma(F(m)\backslash K) = \theta\,\sigma(F(m)).\end{aligned} \tag{12.53}$$

In view of (12.50) we have that the intermediate inequalities must be equalities, which implies (12.48).

This completes the proof of Proposition 12.16.

Remark 12.21 The proof of Proposition 12.16 shows that the weak tangents can be derived using a sequence $\{t_j\}$ which tends to 0. At heart this comes down to Lemma 12.17.

Remark 12.22 (A strengthening of Proposition 12.16) The mappings obtained above (as weak tangents of our original mapping) are not only measure-preserving in their own right, but their own weak tangents all satisfy the same measure-preserving property as well, and with the same value of θ. This can be proved with the same argument as used to show that the limit of the maximizing sequence is measure-preserving.

12.5 Measure-preserving mappings

The measure-preserving property for Lipschitz mappings is pretty strong, if not as strong as bilipschitzness. Let us record the fact that it implies regularity.

Lemma 12.23 (Measure-preserving implies regular) *Let M and N be two Ahlfors regular metric spaces of dimension d, with d-dimensional Ahlfors regular measures μ, ν supported on all of M and N, respectively. Suppose that $f : M \to N$ is Lipschitz and preserves measure in the sense that $\nu(f(K)) = \mu(K)$ for each compact subset K of M. Then f is regular.*

This is an easy consequence of Lemma 12.6, together with the fact that the measures μ and ν are equivalent in size to Hausdorff measure, as in Lemma 1.2.

Notice that we do not really need to assume that ν is supported on all of N, because the image of f must be contained in the support of ν by hypotheses, and because the support of ν is Ahlfors regular.

Regular mappings in general do not distort Hausdorff measure by more than a bounded factor, as in Lemma 12.3, but measure-preserving mappings are better than that, they are almost injective. If f is as in the lemma, then

$$H^d(f(K_1) \cap f(K_2)) = 0 \tag{12.54}$$

whenever K_1, K_2 are disjoint subsets of M.

Let us observe a small converse to this assertion.

Lemma 12.24 *Suppose that $f : M \to N$ is a regular mapping between metric spaces M and N. Suppose that f satisfies (12.54) and that μ is a Borel measure on M which is absolutely continuous with respect to H^d. Let ν be the measure on N obtained by pushing μ forward using f, i.e., $\nu(A) = \mu(f^{-1}(A))$. Then $\nu(f(K)) = \mu(K)$ for any compact subset K of M.*

Of course this is wrong without an assumption like (12.54), because of multiplicities. One can take $f : \mathbf{R} \to \mathbf{R}$ defined by $f(x) = |x|$ for a counterexample.

Let us prove the lemma. Let f, etc., be given as above, and let a compact subset K of M be given in particular. Set $J = f^{-1}(f(K))$. We need to show that $\mu(J\backslash K) = 0$. It suffices to show that $H^d(J\backslash K) = 0$, since we are assuming that μ is absolutely continuous with respect to H^d. To prove this it is enough to show that $H^d(L) = 0$ whenever L is a compact subset of $J\backslash K$, because $J\backslash K$ can be realized as the countable union of compact sets (since J and K are closed).

By assumption, L is disjoint from K, and so $H^d(f(L) \cap f(K)) = 0$ by (12.54). However $f(L) \subseteq f(K)$ by definition, whence $H^d(f(L)) = 0$. This implies that $H^d(L) = 0$, because of Lemma 12.3.

This completes the proof of Lemma 12.24. Of course the real content of the argument is in showing that $H^d(f^{-1}(f(K))\backslash K) = 0$.

Note that the concrete examples of mappings between BPI spaces given in Chapter 11 were measure-preserving.

Let us come back now and think about Proposition 12.16 as a tool for producing measure-preserving mappings. Regularity goes a long way towards the measure-preserving property, and we already knew how to obtain regular weak tangents, as in Proposition 12.8. If we are willing to choose the measures with respect to which a mapping should be measure-preserving, then Lemma 12.24 says that we only need to verify (12.54). This seems a rather weak requirement, since it is implied by injectivity and carries no bounds.

To put this into perspective observe that a Lipschitz mapping between metric spaces (which are doubling and complete) is bilipschitz if and only if it is injective and all of its weak tangents are injective. This is a standard exercise in compactness. As in Remark 12.22, the weak tangents of the mapping provided by Proposition 12.16 have measure-preserving weak tangents. Therefore all of its weak tangents satisfy the weak injectivity property (12.54). This suggests that the mapping provided by Proposition 12.16 behaves better than we have said. We simply lack the language for making this precise.

Another nice point about Proposition 12.16 is that we were able to control to some extent the measures with respect to which we have the measure-preserving property. This fact can be useful in situations with a lot of self-similarity, which could enable us to know the measures exactly.

12.6 Spaces not looking down on each other

Proposition 12.25 *Suppose that M and N are BPI spaces of the same dimension d and that M does not look down on N (Definition 11.1). Let E be a subset* *of M and assume that $f : E \to N$ is regular. Then $f(E)$ is porous (Definition 5.7) in N, and in particular $f(E)$ has measure 0 in N and is even Ahlfors semi-regular of dimension $< d$ (Lemma 5.8).*

Of course Proposition 12.25 is very similar to Proposition 9.20, and the proofs will be similar too. The main point is to prove the following "localized" result.

Lemma 12.26 *Suppose that M and N are BPI spaces of the same dimension d and that M does not look down on N. Then for every $\epsilon > 0$ and $k > 1$ there is a $\delta > 0$ so that if $x \in M$, $0 < r \leq \operatorname{diam} M$, $A \subseteq \overline{B}_M(x,t)$, and $g : A \to N$ is k-Lipschitz, then the image of g is small in the sense that*

$$H^d(\{y \in N : \operatorname{dist}_N(y, g(A)) \leq \delta\, r\}) \leq \epsilon\, r^d. \tag{12.55}$$

Here δ depends on M, N, ϵ, and k, but not on x, r, A, or g.

Thus the failure of looking down implies a quantitative statement.

To prove the lemma we argue by contradiction using compactness. Suppose to the contrary that there is an $\epsilon > 0$ and a $k > 1$ so that for each j we can find $x_j \in M$, $0 < r_j \leq \operatorname{diam} M$, $A_j \subseteq \overline{B}_M(x_j, r_j)$ and a k-Lipschitz mapping $g_j : A_j \to N$ such that

$$H^d(\{y \in N : \operatorname{dist}_N(y, g_j(A_j)) \leq j^{-1}\, r_j\}) \geq \epsilon\, r_j^d. \tag{12.56}$$

We may as well take the A_j's to be closed, because we can always pass to their closure.

Notice that (12.56) implies that $r_j \leq C\,\epsilon^{-1/d} \operatorname{diam} N$. We may rescale the metric on N by a bounded amount if necessary to ensure that $r_j \leq \operatorname{diam} N$ for all j. This will affect k too, but in an acceptable manner.

We want to pass to a limit to get a contradiction from the weak tangents. Choose $a_j \in A_j$ for each j, and set $b_j = g_j(a_j)$. Let M_{a_j,r_j} and N_{b_j,r_j} be the pointed metric spaces defined in the manner of Section 9.1. By passing to a subsequence we may assume that the M_{a_j,r_j}'s and N_{b_j,r_j}'s converge in the sense of Definition 8.9 to weak tangents M' and N' of M and N, respectively. We may also assume that the A_j's converge as subsets of the M_{a_j,r_j}'s to a subset A' of M', as in Definition 8.30 and Lemma 8.31. We can think of $(A_j)_{a_j,r_j}$ as being pointed metric spaces in their own right, which are converging to A' as a pointed metric space (with the same basepoint as M'). We can use the existence of "limits of mapping packages" (Lemma 8.22) to get convergence of $g_j : (A_j)_{a_j,r_j} \to N_{b_j,r_j}$ to the mapping $g' : A' \to N'$. This limiting mapping is also Lipschitz, as in Lemma 8.20.

Note that the A_j's are all contained in the closed ball in M_{a_j,r_j} centered at the basepoint and having radius 2, by definitions. We want to say that the sets $g_j(A_j)$ converge to $g'(A')$ in the sense of Definition 8.30, after passing to a subsequence anyway. We can almost do this using Lemma 8.32, but the statement of this result does not quite apply, because the g_j's are not defined everywhere. This is not a serious issue however, the result extends to this situation. (We omit the details.)

We can also get that the sets

$$\{y \in N : \mathrm{dist}_N(y, g_j(A_j)) \leq j^{-1}\, r_j\} \tag{12.57}$$

converge to $g'(A')$. This is pretty automatic, the point is that these sets are the same as the neighborhoods of size $1/j$ of the images $g_j(A_j)$'s when one uses the rescaled metrics of the N_{b_j,r_j}'s. This small expansion cannot affect the limit of the sets. (We lied here slightly. Since we are always passing to subsequences we ought to incorporate that into the notation, writing j_k everywhere instead of j, but this is a trivial matter.)

Of course the sets $g_j(A_j)$ are also contained in balls in the N_{b_j,r_j}'s of bounded radii about the basepoints, since the g_j's respect the basepoints and are Lipschitz with uniformly bounded norm. Thus we may apply Lemma 8.35 to obtain that

$$H^d(g'(A')) \geq C^{-1}\,\epsilon > 0. \tag{12.58}$$

We are using (12.56) here, of course.

The conclusion now is that M' looks down on N'. This provides a contradiction. We know from Corollary 9.9 that M and M' are BPI equivalent, and similarly for N and N'. In particular M looks down on M', and N' looks down on N. From Lemma 11.5 we conclude that M looks down on N, in contradiction to our initial assumptions. This completes the proof of Lemma 12.26.

It is not difficult to derive Proposition 12.25 from Lemma 12.26. To see this suppose that M and N are as above, and that $f : E \to N$ is regular, $E \subseteq M$. Let B be a ball in N of radius r, so that $f^{-1}(B)$ is covered by a bounded number of balls B_i of the same radius. We apply Lemma 12.26 to obtain that for each $\epsilon > 0$ there is a $\delta > 0$ so that

$$H^d(\{y \in B : \mathrm{dist}_N(y, f(E \cap B_i)) \leq \delta\, r\}) \leq C\,\epsilon\, r^d, \tag{12.59}$$

where C depends on the Lipschitz constant of f but not on B or B_i. Taking the union over the B_i's we get that

$$H^d(\{y \in B : \mathrm{dist}_N(y, f(E)) \leq \delta\, r\}) \leq C\,\epsilon\, r^d, \tag{12.60}$$

where now C depends on the regularity constant for f but not on B. By choosing ϵ small enough we may conclude that $f(E) \cap B$ is thin enough to omit a ball of size $\delta\, r$ contained in B. This implies that $f(E)$ is porous, and the proposition follows.

13

SETS MADE FROM NESTED CUBES

In this chapter we shall look at sets constructed in the manner of Section 2.5. These sets provide a nice "universe" in which to work. They behave well and simply in terms of compactness and weak tangents. We would like to understand the geometric and combinatorial conditions under which there are regular mappings between sets of this type, or bilipschitz mappings or Lipschitz mappings with nontrivial image, but we are far from having definitive answers to these questions. We shall provide some observations about these matters, using the notion of "blocking" sets, and related forms of quantitative topology.

13.1 Preliminary notions

We want to look more at the kinds of sets that can be constructed as in Section 2.5. In particular we want to allow sets that are not constructed by repeating a single rule, so as to have a wider "universe" in which to work.

It will be convenient to have a class of sets which includes the weak tangents of all of its elements. We shall want our constructions to be invariant under the transformations which can take a prescribed cube to the unit cube. This point will be behind some of the definitions which follow.

For this chapter we fix a dimension n and an integer $M \geq 2$. When we speak of "cubes" we mean closed cubes in $\mathbf{R}^n$ with sides parallel to the axes. By "the unit cube" we mean the n-fold product of the unit interval $[0, 1]$. As in Section 2.5 we let $\mathcal{S}$ denote the collection of M^n subcubes of the unit cube obtained by subdividing the unit cube into equal pieces in the obvious manner. We call these cubes the *children* of the unit cube.

Given any cube Q in $\mathbf{R}^n$ we can subdivide it in the same manner to get M^n subcubes of equal size, which we call the children of Q. We do not restrict ourselves to anything like dyadic or "M-adic" cubes here, we intend this notion to be defined for all cubes.

A cube is called a *descendant* of another cube if it arises from any number of generations of passing to children (or if it is the cube itself).

Although our primary interest is in the geometry of a certain class of subsets of $\mathbf{R}^n$, it will be convenient to describe them through classes of collections of cubes.

Definition 13.1 (Normalized families of cubes) *A collection $\mathcal{E}$ of cubes in $\mathbf{R}^n$ is called a normalized family of cubes if it satisfies the following properties: $\mathcal{E}$ contains the unit cube; if $Q \in \mathcal{E}$, then at least one child of Q lies in $\mathcal{E}$, and Q*

is the child of exactly one element of $\mathcal{E}$; each pair of cubes in $\mathcal{E}$ are descended from a common ancestor in $\mathcal{E}$.

Note that a cube Q can be considered as the child of M^n different cubes in $\mathbf{R}^n$. This does not happen in the traditional class of dyadic cubes, and we do not want to allow it in normalized families either.

From a normalized family of cubes we can get a set as follows.

Definition 13.2 (Limiting sets) *Let $\mathcal{E}$ be a normalized family of cubes. Then the* limiting set *E of $\mathcal{E}$ is defined as follows. For each j we take E_j to be the union of the elements of $\mathcal{E}$ of size M^j, and then we set*

$$E = \bigcap_{j=-\infty}^{\infty} E_j \tag{13.1}$$

(Notice that $E_j \subseteq E_{j+1}$ since $\mathcal{E}$ is a normalized family of cubes.)

The limiting set E is closed by definition. It can be described equivalently as the set of points $x \in \mathbf{R}^n$ such that x is contained in elements of $\mathcal{E}$ of arbitrarily small diameter.

Let us record a simple fact.

Lemma 13.3 *If $\mathcal{E}$ is a normalized family of cubes, and $Q_1, Q_2 \in \mathcal{E}$, then either one of Q_1 and Q_2 is contained in the other, or they have disjoint interiors.*

Indeed, the definition of normalized families ensures that Q_1 and Q_2 have a common ancestor. One takes the minimal common ancestor, and either it is one of Q_1 and Q_2, or they are contained in distinct children of this ancestor, in which case they have disjoint interiors.

Next we give a criterion for Ahlfors regularity of the limiting set.

Lemma 13.4 *Suppose that $\mathcal{E}$ is a normalized family of cubes in $\mathbf{R}^n$, and that each element of $\mathcal{E}$ has exactly k children in $\mathcal{E}$ for some integer $k \geq 2$ that does not depend on the cube. Then the limiting set E is Ahlfors regular of dimension d, where $M^d = k$, and the regularity constant is bounded by a constant that depends only on n.*

To see this we argue as in Section 2.5. We can build a nice measure μ on E as follows. For each j we take μ_j to be k^j times the restriction of Lebesgue measure to E_j. One can show that these measures converge weakly to a measure μ on $\mathbf{R}^n$ which is supported on E. The limiting measure μ has the property that

$$k^j \leq \mu(Q) \leq C(n)\, k^j \tag{13.2}$$

whenever Q is an element of $\mathcal{E}$ of size M^j. The constant $C(n)$, which depends only on the dimension, controls the multiplicities which can occur if mass accumulates on a cube Q from the adjacent cubes. (In some cases one can simply take it to be 1.) The lemma follows from Lemma 1.2.

In order to think about the geometry of limiting sets of normalized families of cubes it is helpful to keep track of the *rules* that are used. Let **RULES** denote the collection of all nonempty subsets of $\mathcal{S}$. If $\mathcal{R}$ is an element of **RULES**, a particular "rule", then we define $\mathcal{R}(Q)$ for each cube Q in $\mathbf{R}^n$ to be the collection of children of Q which correspond to elements of $\mathcal{R}$. That is, we can map Q onto the unit cube using a translation and a dilation, and this induces a correspondence between the children of Q and the children of the unit cube. With this correspondence we can pick out the cubes $\mathcal{R}(Q)$ from the children of Q using the rule $\mathcal{R}$.

Given a normalized family of cubes $\mathcal{E}$, to each $Q \in \mathcal{E}$ there is naturally associated a rule $\mathcal{R}_Q \in \mathbf{RULES}$ which is the rule consisting of the children of the unit cube which correspond to the children of Q which lie in $\mathcal{E}$. These rules are like "derivatives" of the normalized family, or "wavelet coefficients" of the limiting set. They code the change in structure from cube to cube.

One can think of parameterizing the set of all normalized families of cubes through families of rules. This is a reasonable idea, but to do it precisely leads to some obnoxious issues of formatting that we shall not address. Nonetheless one should keep the idea in mind. It gives a sense of the range of the set of all normalized families of cubes. Roughly speaking, specifying a normalized family of cubes is like making infinitely many choices of rules which are independent of each other. This infinite collection of choices should not be pictured sequentially, it is more like vertices in a tree.

Let $\mathbf{RULES}(\mathcal{E})$ denote the set of all rules associated to cubes in a normalized family $\mathcal{E}$. If one restricts this set in some way, then one restricts the geometry of $\mathcal{E}$ and the limiting set E. If $\mathbf{RULES}(\mathcal{E})$ has only one element, for instance, then E is either a single point (if the rule has only one element) or a BPI set, as in Section 2.5.

However, $\mathbf{RULES}(\mathcal{E})$ could have more than one element and the limiting set E still be BPI. For instance, if all the rules in $\mathbf{RULES}(\mathcal{E})$ have the same number of elements, and if each rule $\mathcal{R}$ in $\mathbf{RULES}(\mathcal{E})$ has the property that distinct elements of $\mathcal{R}$ are disjoint, then the limiting set E is BPI, and is in fact bilipschitz equivalent to an unbounded version of a Cantor set whose geometry depends only on the number of elements in the rules in $\mathbf{RULES}(\mathcal{E})$.

In general the interrelationships between the cubes inside the rules matter more than their particular positions. There are some subtleties in the way that the rules reflect the geometry of the limiting sets, as in Section 11.6.

13.2 Convergence of families of cubes

Definition 13.5 (Convergence of families of cubes) *Let $\{\mathcal{E}_j\}$ be a sequence of sets of cubes in $\mathbf{R}^n$, and let $\mathcal{E}$ be another set of cubes. We say that $\{\mathcal{E}_j\}$ converges to $\mathcal{E}$ if each cube Q in $\mathcal{E}$ lies in all but finitely many $\mathcal{E}_j$'s and each cube Q not in $\mathcal{E}$ lies in only finitely many $\mathcal{E}_j$'s.*

This corresponds to the most classical version of "convergence of sets", and it has nothing to do with families of cubes *per se*.

Lemma 13.6 *If $\{\mathcal{E}_j\}$ is a sequence of normalized families of cubes in $\mathbf{R}^n$, then there is a subsequence which converges to another normalized family of cubes $\mathcal{E}$. Moreover, every element of* **RULES**$(\mathcal{E})$ *lies in* **RULES**$(\mathcal{E}_j)$ *for all but finitely many choices of j.*

The main point here is that normalized families of cubes force a kind of countability that lends itself to a Cantor diagonalization procedure.

Let $\mathcal{Q}$ denote the collection of all cubes Q in $\mathbf{R}^n$ which are descended from a cube Q' which is itself an ancestor of the unit cube. One can think of $\mathcal{Q}$ as the set of all cousins of the unit cube.

There are only countably many elements of $\mathcal{Q}$. This is easy to check, because there are only countably many ancestors of the unit cube. Keep in mind that the unit cube is a child of M^n different cubes, though.

Normalized families of cubes are always contained in $\mathcal{Q}$. To check the convergence of a normalized family one need only think about cubes in $\mathcal{Q}$.

Let $\{\mathcal{E}_j\}$ be a sequence of normalized families of cubes, as above. Given a cube $Q \in \mathcal{Q}$, we can pass to a subsequence to reduce to the case where Q lies either in all but finitely many of the $\mathcal{E}_j$'s or in only finitely many of them. Using a Cantor diagonalization argument we can find a subsequence where this is true for all $Q \in \mathcal{Q}$. We take $\mathcal{E}$ to be the set of $Q \in \mathcal{Q}$ which lie in $\mathcal{E}_j$ for all but finitely many j's. It is easy to see that $\{\mathcal{E}_j\}$ converges to $\mathcal{E}$, using the definitions.

It is also easy to check that $\mathcal{E}$ is a normalized family of cubes, since the $\mathcal{E}_j$'s are, just using the definitions. It is also easy to check that each rule that occurs within $\mathcal{E}$ occurs also in all but finitely many of the $\mathcal{E}_j$'s. (Of course one can say much more like this.)

This completes the proof of the lemma.

Lemma 13.7 *Suppose that $\{\mathcal{E}_j\}$ is a sequence of normalized families of cubes which converges to another normalized family of cubes $\mathcal{E}$. Let E_j, E be the corresponding limiting sets. Then $\{E_j\}$ converges to E in the sense of Definition 8.1.*

This is easy to check. The main points are these. If $\mathcal{F}$ is a normalized family of cubes, and if $Q \in \mathcal{F}$, then the limiting set F has elements in Q. This is because every element of $\mathcal{F}$ has at least one child in $\mathcal{F}$. Conversely, if $x \in F$ and j is any integer, then there must be a cube $Q \in \mathcal{F}$ of size M^j which contains x. Again this uses the definition of a normalized family to ensure that the parent of any element of $\mathcal{F}$ also lies in $\mathcal{F}$. Using this simple correspondence between points and cubes it is easy to check that convergence of normalized families of cubes implies convergence of the limiting sets. This proves the lemma.

13.3 Weak tangents of families of cubes

Let $\mathcal{E}$ be a normalized family of cubes in $\mathbf{R}^n$. Given a cube $Q \in \mathcal{E}$, let $\mathcal{E}_Q$ be the family of cubes that is obtained as follows. First we take the mapping from $\mathbf{R}^n$ to itself which is a composition of a translation and a dilation and which sends

Q onto the unit cube. Then we take $\mathcal{E}_Q$ to be the collection of cubes which are the images of elements of $\mathcal{E}$ under this mapping. In other words, we move $\mathcal{E}$ by translation and dilation so that Q becomes the unit cube. It is easy to see that $\mathcal{E}_Q$ is also a normalized family of cubes.

Let $WT(\mathcal{E})$ denote the collection of all normalized families of cubes which arise as limits of the $\mathcal{E}_Q$'s, $Q \in \mathcal{E}$, in the sense of Definition 13.5. (This includes the $\mathcal{E}_Q$'s themselves, by choosing constant sequences of Q's.) This is like the previous constructions of weak tangents. We want to check that this notion of weak tangents is compatible with the previous one for sets.

Lemma 13.8 *Suppose that $\mathcal{E}$ is a normalized family of cubes in $\mathbf{R}^n$, and let E be the limiting set of $\mathcal{E}$. Let $\mathcal{E}'$ be an element of $WT(\mathcal{E})$, and let E' be the corresponding limiting set. Then for each point p in the intersection of E' with the unit cube we have that the pointed metric space $(E', |x-y|, p)$ is a weak tangent of E (up to isometric equivalence, if you wish).*

By assumption there is a sequence of cubes $\{Q_j\}$ in $\mathcal{E}$ such that $\mathcal{E}'$ is the limit of $\mathcal{E}_j = \mathcal{E}_{Q_j}$. Let E_j denote the limiting sets of the $\mathcal{E}_j$'s. We know from Lemma 13.7 that $\{E_j\}$ converges to E' in the sense of Definition 8.1. This implies that we can find points $q_j \in E_j$ such that $\{q_j\}$ converges to p.

Let F' be the image of E' under the mapping $x \mapsto x - p$, and let F_j denote the image of E_j under the mapping $x \mapsto x - q_j$. It is easy to check that F_j converges to F' in the sense of Definition 8.1, because of the corresponding convergences of the E_j's and the q_j's.

There is a point $p_j \in E$ which corresponds to q_j. That is, $\mathcal{E}_j$ is the image of $\mathcal{E}$ under the translation and dilation which sends Q to the unit cube, E_j is the image of E under the same mapping, and there is a corresponding point $p_j \in E$.

Thus the F_j's are really just translations and dilations of E, normalized in such a way that the origin lies in each F_j. Therefore their limit F' is a weak tangent to E, viewed as a pointed metric space with the origin as basepoint. Of course F' is the same as E' except for a translation, and the lemma follows.

Next we check the converse.

Lemma 13.9 *Suppose that $\mathcal{E}$ is a normalized family of cubes in $\mathbf{R}^n$, and let E be the limiting set of $\mathcal{E}$. Let $(M, d(\cdot,\cdot), p)$ be a pointed metric space which is a weak tangent to E (as in Section 9.1). Then there is a normalized family of cubes $\mathcal{E}' \in WT(\mathcal{E})$ with limiting set E', a point q in the intersection of E' with the unit cube, and a scale factor $r \in [1, M]$ such that $(M, d(\cdot,\cdot), p)$ is isometrically equivalent to $(E', r\,|x-y|, q)$.*

Indeed, by definitions $(M, d(\cdot,\cdot), p)$ can be realized as the limit of the pointed metric spaces $(E, t_j^{-1}\,|x-y|, p_j)$ for some sequences $\{p_j\} \subseteq E$ and $\{t_j\} \subseteq (0,\infty)$. We do not need to work with the abstract definition of convergence of metric spaces here, we can simply use convergence of subsets of $\mathbf{R}^n$, as observed in Section 9.1. (That is, one gets the same spaces up to isometric equivalence.) Thus if we take F_j to be the subset of $\mathbf{R}^n$ which is the image of E under the affine mapping $x \mapsto t_j^{-1}(x - p_j)$, then we may assume that $\{F_j\}$ converges in the sense of Definition 8.1 to a subset F of $\mathbf{R}^n$, in such a way that $(F, |x-y|, 0)$ is isometrically equivalent to our original weak tangent $(M, d(\cdot,\cdot), p)$ as pointed metric spaces. (Strictly speaking we have passed to a subsequence here.)

For each j we can find a number $r_j \in [1, M]$ and integer $l(j)$ such that $t_j = r_j\, M^{l(j)}$.

Choose for each j a cube $Q_j \in \mathcal{E}$ of sidelength $M^{l(j)}$ which contains $p_j \in E$. Set $\mathcal{E}_j = \mathcal{E}_{Q_j}$. By passing to a subsequence we may assume that $\{\mathcal{E}_j\}$ converges to a normalized family of cubes $\mathcal{E}'$. Let E' be the limiting set of $\mathcal{E}'$, and let E_j be the limiting set of $\mathcal{E}_j$. Thus E_j is the image of E under the translation and dilation which send Q_j to the unit cube. Let q_j denote the image of p_j under this same mapping, so that q_j lies in the intersection of E_j with the unit cube.

We have that E_j is the image of F_j under the mapping $x \mapsto r_j\,(x + q_j)$. Indeed, this is the combination of translation and dilation which sends the origin to q_j and converts the dialation factor t_j^{-1} into the dilation factor $M^{-l(j)}$, which are correct for E_j.

By passing to a subsequence we may assume that $\{q_j\}$ converges to a point q in the unit cube and that $\{r_j\}$ converges to some $r \in [1, M]$. In this case the E_j's will converge to the image of F under $x \mapsto r\,(x + q)$.

We know from Lemma 13.7 that the E_j's converge to E' in the sense of Definition 8.1. Thus E' is the image of F under $x \mapsto r\,(x+q)$. This says exactly that $(F, |x-y|, 0)$ is isometrically equivalent to $(E', r\,|x-y|, q)$, and the lemma follows.

Let us mention also the following.

Lemma 13.10 *Let $\mathcal{E}$ be a normalized family of cubes in $\mathbf{R}^n$, and let $\mathcal{E}'$ be an element of $WT(\mathcal{E})$. Then* **RULES**$(\mathcal{E}') \subseteq$ **RULES**$(\mathcal{E})$.

This is an easy consequence of Lemma 13.6.

13.4 Mappings

Proposition 13.11 *Let $\mathcal{E}$ and $\mathcal{F}$ be normalized families of cubes in $\mathbf{R}^n$, with limiting sets E and F. Suppose that E is Ahlfors regular of dimension d, and that there exists a compact set $Z \subseteq E$ and a Lipschitz mapping $g : Z \to F$ with $H^d(g(Z)) > 0$. Then there exists $\mathcal{E}' \in WT(\mathcal{E})$ and $\mathcal{F}' \in WT(\mathcal{F})$ with limiting sets E', F' and a regular mapping (Definition 12.1) from E' into F'.*

The idea here is that the existence of the "substantial" mapping g ought to imply some kind of combinatorial compatibility between $\mathcal{E}$ and $\mathcal{F}$. We do not achieve that here, we are simply getting an improvement in the kind of mapping. This improvement comes from Proposition 12.8 and Lemma 13.9. We could also use Proposition 12.16 to get better mappings under suitable conditions.

In the next sections we discuss geometric consequences of the existence of regular mappings, and we consider examples from limiting sets of normalized families of cubes.

13.5 Subsets that block connectedness

Definition 13.12 *Let $(N, d(x, y))$ be a metric space, let E be a subset of N, and let s be a positive number. We say that E is* s-blocking *in N if every pair of points $z, w \in N \backslash E$ with $d(z, w) > s$ are disconnected by E in N (i.e., there is no connected subset of $N \backslash E$ which contains both z and w).*

The idea is that the size of the blocking set says something about the geometry of N. It provides a measurement of the number of curves in the set, and more generally the number of continua. (We could make more precise definitions aimed at curves, or rectifiable curves, etc.) Let us illustrate this idea with some examples.

Examples 13.13 (a) If N is totally disconnected, like a Cantor set, then the empty set is s-blocking for all $s > 0$.

(b) If N is the unit interval $[0, 1]$, and if $E \subseteq [0, 1]$ is s-blocking, then E should have at least $s^{-1} - 1$ elements.

(c) If N is the unit cube in $\mathbf{R}^n$ and E is s-blocking with $s < \frac{1}{2}$, say, then $H^{n-1}(E) \geq C(n)\, s^{-1}$, where $C(n)$ does not depend on s. (The intersection of E with any segment in the cube must be s-blocking in that segment.)

In all of these cases one can easily construct s-blocking sets of the same approximate size as in the bound.

Suppose that $N = K \times [0, 1]$ for some space K. Think of K as being a Cantor set for instance. If $E \subset N$ is s-blocking, then for each $p \in K$ $E \cap (\{p\} \times [0, 1])$ should be s-blocking in $[0, 1]$ and therefore contain at least $s^{-1} - 1$ elements. This permits one to read off conditions on the size of E in terms of the size of K. For instance, E should have Hausdorff measure at least $s^{-1} - 1$ times that of K in any dimension, and in particular the Hausdorff dimension of E should be at least as big as the Hausdorff dimension of K when $s < 1$. If K is totally disconnected these bounds can be achieved easily by chopping up $[0, 1]$ into pieces.

Of course one can give general results about the relationship between the blocking sets of a Cartesian product and its individual factors.

Note that one can get general upper bounds on the size of the smallest blocking set which are like those for Euclidean spaces, as in the following for instance.

Lemma 13.14 *Suppose that $(N, d(x, y))$ is a metric space which is Ahlfors regular of dimension n and which has diameter ≤ 1. Then for each $s \in (0, 1/2)$ we can find an s-blocking set $E \subseteq N$ such that $H^{n-1}(E) \leq C\, s^{-1}$, where C depends on the regularity constants but not on s. If we drop the assumption of Ahlfors regularity and ask only that N be compact and that $H^n(N) < \infty$, then we can find s-blocking sets E with $H^{n-1}(E) < \infty$ for all $s > 0$.*

The proof of this is quite straightforward. Suppose first that N is regular with dimension n and has diameter ≤ 1. This implies that we can find a collection A of points in N such that

$$N \subseteq \bigcup_{a \in A} B(a, s/4) \tag{13.3}$$

and A has at most $C\,s^{-n}$ elements. (See Lemma 5.1.) Set $\Sigma(x,r) = \{y \in N : d(x,y) = r\}$. For each $a \in A$ we can find a radius $r(a) \in (s/4, s/2)$ such that

$$H^{n-1}(\Sigma(a, r(a))) \leq C\,s^{-1}\,H^n(B(a, s/2)) \leq C\,s^{n-1}. \tag{13.4}$$

This follows from the co-area theorem applied to the Lipschitz mapping $x \mapsto d(x,a)$. (See Theorem 2.10.25 on p.188 of [Fe], Theorem 7.7 in [Ma], or Proposition 14.1 in [Se6].) Take $E = \bigcup_{a\in A} \Sigma(a, r(a))$. It is easy to see that this is s-blocking; if $z, w \in N\backslash E$ satisfy $d(z,w) > s$, then $z \in B(a, s/4)$ for some $a \in A$ but then $d(w,a) > s/2$ for that same a. Any connected set containing z and w must intersect $\Sigma(a, r(a))$, and hence E. Thus E is s-blocking. Our bounds on the number of elements of A and on the mass of the spheres $\Sigma(a, r(a))$ implies that $H^{n-1}(E) \leq C\,s^{-1}$, as desired.

If we only know that N is compact and has finite n-dimensional Hausdorff measure, then we can repeat the argument, but without the precise bounds on the number of elements of A (now it is simply finite) or on the mass of the spheres $\Sigma(a, r(a))$, but we still get that $H^{n-1}(E) < \infty$.

Of course there are many variations on this theme.

Thus we have some universal upper bounds for the minimal blocking sets which are approximately achieved for Euclidean cubes and more generally spaces of the form $N = K \times [0,1]$. (If one is interested in precise constants then cubes have to be treated differently, their constants will be larger than for the product of $[0,1]$ with a Cantor set for instance, because one must block connections in more directions.)

On the other hand the minimal size of the blocking sets might be much smaller in particular cases. For Cantor sets the empty set is blocking. Let us look now at some other examples.

Consider the snowflake $([0,1], |x-y|^\alpha)$, which is Ahlfors regular of dimension $1/\alpha$. A minimal s-blocking set should have at least $s^{-\alpha} - 1$ elements, and this bound can be achieved. (More generally E is s-blocking for $(N, d(x,y))$ if and only if E is s^α-blocking for $(N, d(x,y)^\alpha)$.)

Suppose that $(N, d(x,y))$ enjoys the *local linear connectedness* property that for each $z, w \in M$ there is a connected set which contains them which has diameter $\leq C\,d(z,w)$. Suppose that $E \subseteq N$ is s-blocking, and let us make the normalizing assumptions that $s < 1 \leq \operatorname{diam} N$. Then every ball of radius $C\,s$ must contain an element of E for a constant C which depends only on the linear local connectedness constant of N and not on s. If N is Ahlfors regular of dimension γ, then E must contain at least $C^{-1}\,s^{-\gamma}$ elements. This is the correct approximate bound for the snowflake curves $([0,1], |x-y|^\alpha)$. It is also the correct approximate bound for the examples of the fractal tree and Sierpinski gasket discussed in Section 2.4.

The fractal tree and the Sierpinski gasket are very different from the snowflake curve. These examples point out the relevance of other notions of blocking sets, with restricted types of connections. One can confine oneself to connections between points provided by rectifiable curves, for instance. More generally one can impose bounds on the "size" of the connecting sets. If one restricts oneself to rectifiable curves, for instance, then snowflakes behave like totally disconnected sets – there are no nontrivial rectifiable curves – while the fractal tree and Sierpinski carpet behave in the same manner as before. Indeed, for these sets any pair of points can be connected by a rectifiable curve whose length (instead of diameter) is bounded by a multiple of their distance.

One can also make many variations on these concepts as well.

The *Sierpinski carpet* described in Section 2.4 has a more complicated structure for the notion of blocking sets than the examples mentioned above. Let K denote the Sierpinski carpet, and let A denote the standard middle-thirds Cantor set. Thus K contains a copy of $A \times [0,1]$. This leads to some lower bounds on the size of blocking sets, but they are not the right ones.

In the construction of K one gets for each positive integer j a collection $\mathcal{Q}_j$ of 8^j squares of sidelength 3^{-j}. K is contained in the union of these 8^j squares, and the intersection of K with any one of these squares Q looks like K again, i.e., $K \cap Q = \tau(K)$, where $\tau : \mathbf{R}^2 \to \mathbf{R}^2$ is the combination of a translation and dilation (by 3^{-j}) which takes the unit square to Q.

Let A_j be a copy of A but shrunk to the size of 3^{-j}. Then each $K \cap Q$, $Q \in \mathcal{Q}_j$ contains a translation of $A_j \times [0, 3^{-1}]$. (It also contains another copy with the axes flipped.)

Suppose that $E \subseteq K$ is s-blocking for some $s < 3^{-j}$. Then E has to "block" each little copy of $\{a\} \times (0, 3^{-j})$ inside $K \cap Q$ for each $Q \in \mathcal{Q}_j$ and each $a \in A_j$. More precisely, the projection of $E \cap Q$ onto the first axis has to contain the whole copy of A_j there. This means that in size E should be at least as big as 8^j copies of A_j. (To really get 8^j, rather than a uniform fraction of it, one should be careful not to count twice the same piece of E from adjacent squares. This is actually excluded by using $(0, 3^{-1})$ above instead of a closed interval, except for a finite number of points. We do not really care about this though, even being very sloppy one would still get a uniform fraction of 8^j.)

For instance, if the dimension δ is chosen so that $3^\delta = 2$, then $H^\delta(A_j) \sim 2^{-j}$, and the preceding argument leads to the bound $H^\delta(E) \geq C^{-1}\, 4^j$.

This analysis is pretty sharp, for the following reason. Consider the set

$$F = (A \times \{\tfrac{1}{2}\}) \cup (\{\tfrac{1}{2}\} \times A). \tag{13.5}$$

One can check that

$$K \cap \{([0,1] \times \{\tfrac{1}{2}\}) \cup (\{\tfrac{1}{2}\} \times [0,1])\} = F. \tag{13.6}$$

This follows from the construction of K in Section 2.4. Thus F is sort of a "basic blocking set", one cannot connect a pair of points in K without touching F if the points lie in distinct quarters of the unit square.

Now let F_j denote the union of the copies of F inside each $Q \in \mathcal{Q}_j$. Thus F_j is the union of $2 \cdot 8^j$ copies of A_j, and in particular $H^\delta(F_j) \leq C\, 4^j$. One can check that F_j is s-blocking in K with $s = \sqrt{2} \cdot 3^{-j}$. More precisely, if two points in $K\backslash F$ lie in the same connected component of $K\backslash F$, then they have to lie in a square of size 3^{-j}. (This square will not lie in $\mathcal{Q}_j$, but it will be centered on a vertex of one of those squares. The point is that F_j puts walls in K in the middle of these squares, but it does not prevent points from travelling across the sides of these squares.)

Thus our upper bounds and lower bounds approximately match for the Sierpinski carpet. (They differ by bounded factors.)

These examples illustrate how the size of blocking sets reflect the geometry of a set, and to some extent the combinatorics of the family cubes when the set arises as the limiting set of a family of cubes.

The concept of blocking set (and its variations) obviously behaves well under bilipschitz equivalence, but regular mappings cooperate well with it too.

Lemma 13.15 *Suppose that $(N_1, d_1(\cdot,\cdot))$ and $(N_2, d_2(\cdot,\cdot))$ are metric spaces, and that $f : N_1 \to N_2$ is regular. Suppose also that E_2 is s-blocking in N_2 for some $s > 0$. Then $E_1 = f^{-1}(E_2)$ is $C\,s$-blocking in N_1, where C depends only on the regularity constant of f.*

To understand this it is helpful to express the blocking condition in slightly different words: we want to say that $N_1\backslash E_1$ contains no connected subsets of diameter $> C\,s$. Thus all that we really need to know is that

$$\operatorname{diam}_{N_2} f(\Gamma) \geq C^{-1} \operatorname{diam}_{N_1} \Gamma \tag{13.7}$$

whenever $\Gamma \subseteq N_1$ is connected. In the case of regular mappings this property is provided by Lemma 12.4.

Regular mappings behave much better than this, and a large measure of their good behavior is reflected by the fact that the sets E_1 and E_2 have to have approximately the same size. They can be very different geometrically – E_2 might be an interval and E_1 might be a Cantor set – but in terms of size they have to be approximately the same, where "size" is anything that can be measured in terms of coverings of balls. This follows from the definition of regularity, and it includes Hausdorff measure, for instance, as in Lemma 12.3.

In short, regular mappings respect well the size of blocking sets.

We can use this to obtain geometric restrictions on the existence of regular mappings between spaces. This is another version of the idea that spaces can become more connected under a mapping but not less. The examples above provide concrete illustrations of this principle.

We can then go back to Proposition 12.8 to get restrictions on the existence of even partially defined Lipschitz mappings which are not too degenerate.

13.6 Going further

This notion of blocking sets can be a useful tool but it is also crude. Consider the Sierpinski gasket and the fractal tree, for instance. They are obviously quite different because they have different dimensions, but let us ignore that for the moment and pay more attention to the combinatorial geometry.

In terms of blocking sets the two are very similar. In each case there is a blocking set of approximately optimal size which is finite and whose growth one can compute easily. Nonetheless the two sets have fundamental differences in structure. The fractal tree is much more "primitive". The elements in the most obvious blocking sets for these two fractals are connected to each other in very different ways. The connections for the tree are simpler.

We can make this idea precise by associating finite graphs to these sets in the following manner. Denote the Sierpinski gasket by G and the fractal tree by F. The construction of G and F provides sequences of sets $\{G_j\}$ and $\{T_j\}$, where each G_j is the union of 3^j triangles (with interiors included) of sidelength 3^{-j}, each F_j is the union of 5^j squares of sidelength 3^{-j}, and $\{G_j\}$ and $\{T_j\}$ are decreasing sequences of sets, with $\cap_j G_j = G$, $\cap_j T_j = T$.

We would like at first to define finite graphs associated to G_j and F_j whose edges correspond to the constituent triangles or squares and whose vertices correspond to points of intersection for them. This does not quite work, though, having edges correspond to squares or triangles. Sometimes the edges would need more than two endpoints! To fix this we need to use more edges.

Given a fixed triangle T in the plane, we can associate to it the graph Y consisting of the vertices and centerpoint of T as vertices, and the line segments from the three vertices of T to the centerpoint of T as the edges of Y. Similarly, given a square S in the plane, we associate to it a graph X whose vertices are the four vertices of S together with the centerpoint of S, and whose edges are the line segments which connect the vertices of S to the centerpoint. (In this discussion we consider "triangles" and "squares" to be regions in the plane with their boundaries, and not simply the boundaries themselves.) Thus Y and X look approximately like the letters used to denote them.

We can now associate graphs $Y(G_j)$ and $X(F_j)$ to G_j and F_j by replacing each constituent triangle T in G_j (square S in F_j) with the corresponding Y (X), and then taking the union over all the constituent triangles in G_j (squares in F_j). The idea is that $Y(G_j)$ and $X(F_j)$ make good approximations to G_j and F_j. Not merely in the sense of (Hausdorff) distance, but also with respect to combinatorics and topology.

To make this precise, let us define mappings $g_j : G_j \to Y(G_j)$ and $f_j : F_j \to X(F_j)$ in the following manner. Given a triangle T we can map T to $Y(T)$ in such a way that the three vertices of T are held fixed and no other elements of T are mapped into these points. Similarly we can map a square S to $X(S)$ in such a way that the vertices of S are held fixed and no other elements of S are mapped to these four points. Of course we want these mappings to be continuous, and we can even take them to be piecewise-linear. We can define g_j and f_j as above simply by combining copies of these building blocks on the constituent pieces. Because the triangles in G_j and the squares in F_j intersect only in vertices, and never in edges, the mappings g_j and f_j are also continuous. In fact one can check that they are Lipschitz with bounded constant (independently of j).

These mappings capture information about blocking sets. If we let V denote the totality of the vertices of the triangles in G_j, then V is an s-blocking set for $Y(G_j)$ with a value s comparable to 3^{-j}. The mapping g_j ensures that V is also a blocking set for G_j with a similar value of s, and hence a blocking set for G itself. We are using here the fact that $V = g_j^{-1}(V)$, by construction. Similarly, if W is the set of vertices of the squares in F_j, then W is a blocking set for $X(F_j)$, F_j, and F with a value of s of about 3^{-j}. (One can be more efficient here and not use vertices of triangles or squares that are not contained in more than one triangle or square.)

Thus we can "embed" the information about blocking sets for G and F into this story of mappings for which there is more structure.

We also have "dual" mappings $h_j : Y(G_j) \to G$ and $e_j : X(F_j) \to F$. It is easier to get maps into G_j and F_j, but let us be more careful and get maps into the limiting fractals G and F. We define these mappings first on the individual pieces. If T is one of the triangles in G, then we want to map the associated graph Y into $T \cap G$ in such a way that the three vertices of T are held fixed. This is not hard to do (but do not worry about making this mapping injective.) One can combine these pieces to get $h_j : Y(G_j) \to G$ which is continuous, fixes the elements of V, and maps $Y(G_j) \cap T$ into $G \cap T$ for each of the constituent triangles T in G_j. One defines e_j in exactly the same manner. Given a square S in F_j, we first map the corresponding graph X into $S \cap F$ in such a way that the vertices of S are held fixed. Again it is not hard to see that this is possible. Then we combine the pieces to get a continuous mapping $e_j : X(F_j) \to F$ which fixes all the elements of W and maps $S \cap X(F_j)$ into $S \cap F$ for each of the constituent squares S of F_j. These mappings can even be taken to be Lipschitz with a uniform bound. This comes down to the fact that we can take the building block mappings to be Lipschitz, and the fact that the building blocks can be taken to be all the same modulo translations and dilations.

Thus the combinatorial aspects of the construction of the fractal sets G and F can be re-embedded into the sets in a nontrivial manner. These re-embeddings help to code some of the combinatorics in such a way as to make them visible through more primitive considerations of the geometry of G and F, without regard to the particular manner in which we chose to construct them. For example, if we have a Lipschitz mapping from some space M into G or F, then we get a Lipschitz mapping from M into each of the corresponding $Y(G_j)$'s or $X(F_j)$'s, with uniform bounds. Information about the original mapping – like regularity or bilipschitzness – leads to information about M, in terms of blocking sets for instance, but in fact with these mappings we get quite a bit more than that.

Similarly, if we have a Lipschitz mapping from G or F into some space N, then we have mappings from the corresponding $Y(G_j)$'s or $X(F_j)$'s into N. Information about the original mapping leads to information about these induced mappings.

We mentioned before that the story of blocking sets alone does not provide a good way to distinguish the combinatorial geometries of G and F, setting aside the cruder issue of raw dimension for this discussion. We said that the geometry of the fractal tree F should be more primitive than that of the Sierpinski gasket G. This story of approximating graphs helps to make this more precise. Mappings between such fractals lead to mappings between the associated graphs. The graphs associated to the fractal tree are trees, the graphs associated to the Sierpinski carpets are not. It is easier to map trees into other graphs in a nontrivial way (i.e., with finite multiplicities, the combinatorial counterpart of regular mappings) than the other way around. For the trees it can often be simply a question of adding connections, maintaining injectivity on the interiors of the edges. For mappings into trees one is likely to be forced to retrace the edges, perhaps many times.

The general idea here is to look for ways in which the "transcendental" geometry of fractals like G and F can be coded through finite combinatorics. We construct these fractals through combinatorial means, we get sets with certain geometry in terms of metric and measure, and we would like to recapture the combinatorics from the metric and measure as much as possible.

We can put these ideas in a broader framework using "topological dimension theory" [HW]. Instead of starting with the definition of topological dimension, let us use the following characterization of Alexandroff: a compact metric space X has topological dimension $\leq n$ if and only if for each $\epsilon > 0$ there is a continuous mapping f from X into a finite polyhedron P of dimension $\leq n$ (a finite union of simplices of dimension $\leq n$) such that $\operatorname{diam} f^{-1}(p) \leq \epsilon$ for every $p \in P$. (See the corollary on p.72 of [HW], and also Definition V 4 on p.57 of [HW].) Topological dimension theory in general provides many characterizations and criteria for this property, as in [HW].

One often hears "fractals" described as sets for which the Hausdorff dimension is strictly larger than the topological dimension. (The only other possibility is equality.) This is a nice enough idea, but it does not provide a great deal of information.

One can try to get more information by adding structure to the notion of topological dimension. One can start with the idea of approximating a space by a polyhedron as above, for instance, and then impose Lipschitz bounds on the mapping, or bounds on the complexity of the polyhedron, or bounds on the fibers of the mappings, etc., as in the earlier discussion of the Sierpinski gasket and fractal tree.

One can attempt similar analyses for other types of fractals. Using approximating mappings one can get plenty of extensions and refinements of the idea of blocking sets, for instance. One can try to account for larger families of curves, as in the Sierpinski carpet, by observing that there should not be points in the image with only one preimage, as there were for the Sierpinski gasket and fractal tree. One can look at lower bounds for the sizes of these inverse images, as in the story of blocking sets.

One can also try to account for snowflake behavior in intelligent ways.

The bottom line is really that there are plenty of things to try. At the moment the situation is somewhat like that of a zoo under construction.

See also [Se6] for stories about quantitative topology and the distribution of curves in a set.

The general idea of looking at the distribution of curves in sets like these may well be better than the particular suggestions that we have offered here for doing this. One should pay more attention to the manner in which the curves interact, at how they connect from one particular place to another. There is a big difference between the Sierpinski carpet and the product of an interval with a Cantor set, for instance. The manner in which the curves are tangled together is also important, and can make it difficult to map the curves into another space without having a lot of crossings.

There is an amusing "opposite" to the questions discussed here. Instead of looking for geometric properties that regular mappings might preserve or affect in a controlled way, one can start with a space and try to make another space with stronger or weaker geometric properties (in terms of connectedness, for instance) and a regular mapping between the two spaces. Under what conditions is a space the image under a regular mapping of something like a fractal tree, or a Sierpinski gasket, or a Sierpinski carpet? These are natural questions, and Section 15.5 provides a type of construction which can be used as a basis for approaching the issue.

14

BIG PIECES OF BILIPSCHITZ MAPPINGS

14.1 Introduction

Definition 14.1 *Let $(M, d(\cdot,\cdot))$ and $(N, \rho(\cdot,\cdot))$ be metric spaces, and suppose that $f : M \to N$ is Lipschitz. We say that f has a* big bilipschitz piece *if there is a closed subset E of M such that the restriction of f to E is bilipschitz and if $H^{\kappa}(E) > 0$, where κ is the Hausdorff dimension of M.*

Of course the restriction to closed sets comes for free, since a Lipschitz mapping which is bilipschitz on a subset is also bilipschitz on the closure.

Think of the case where M is Ahlfors regular, so that Hausdorff measure behaves well.

Definition 14.1 should not be confused with our earlier Definition 7.20, in which we asked for big bilipschitz pieces at all scales and locations, with uniform bounds.

It is a classical fact that a Lipschitz mapping between Euclidean spaces has a big bilipschitz piece as soon as its image has positive measure in the correct Hausdorff dimension. (See [Fe].) The main point is that any such Lipschitz mapping can be modified on a set of arbitrarily small measure to get a C^1 mapping. For C^1 mappings one can get big Lipschitz pieces at places where the Jacobian is nonvanishing, because of the inverse function theorem.

We have seen that this fact extends to the case where the target space is any metric space (Theorem 11.12). There are quantitative results on the existence of big bilipschitz pieces in [D3, J2], and these results are important for studying uniformly rectifiable sets.

On the other hand there are plenty of spaces on which we can make Lipschitz mappings into Euclidean spaces with big image but no big bilipschitz piece. Cantor sets and snowflake curves can look down on Euclidean spaces without having a big bilipschitz piece, for instance. For these spaces there is nothing like the differentiability almost everywhere of Lipschitz functions.

One can view this absence of rigidity as a reflection of the fractal nature of the domain spaces, the lack of such good structure as Euclidean spaces enjoy. Alternatively one could say that it is too easy to build mappings into Euclidean spaces. That Euclidean spaces have a very rigid structure in terms of mappings defined on them, but very little rigidity in terms of mappings into them. Cantor sets are the opposite; they have little rigidity in terms of mappings on them, but a lot of rigidity in terms of mappings into them. Snowflake curves are somewhere in between.

Notice that any Lipschitz mapping from $([0,1], |x-y|^s)$ into itself with big image does have a big bilipschitz piece, because this question is equivalent to its counterpart for $[0,1]$ with the standard metric.

One might hope for positive results in general if one restricts one's attention to mappings from a space to itself. This is an appealing idea, that what is bad for the image is the opposite of what is bad for the domain, so that a balance is attained. In this chapter we shall give examples to show that this is too optimistic. We shall see that there are Lipschitz mappings from a Cantor set into itself with good properties but no big bilipschitz piece.

Instead of asking for a big bilipschitz piece we could look for the existence of a bilipschitz weak tangent. If the domain space is Ahlfors regular then the presence of a big bilipschitz piece implies the existence of a bilipschitz weak tangent. This is not hard to check, by blowing up at a point of density and using Lemmas 9.12 and 9.13. The existence of a bilipschitz weak tangent is sufficient to conclude BPI equivalence if the two metric spaces are BPI spaces of the same dimension, because of Proposition 7.1, Corollary 9.9, and Proposition 7.5.

We shall see, however, that there are Lipschitz mappings from a Cantor set onto itself without any bilipschitz weak tangents.

It remains unclear as to exactly what positive results to look for. Cantor sets are certainly much flabbier than fractals like Sierpinski carpet, the Sierpinski gasket, and the fractal tree, which should enjoy more interesting rigidity properties. Mappings from these spaces into themselves have to respect the complicated structure of "connections" within. This should lead to strong restrictions on the mappings (although not necessarily big bilipschitz pieces).

This chapter is very closely tied to Chapter 11. In order to relate the looking-down property to BPI equivalence it would be very helpful to have conditions under which Lipschitz mappings with big image necessarily have big bilipschitz pieces. We know that this is not true in general, but one could hope to have some natural criteria. In this chapter we see that we should not hope for too much in general.

Still we do not know whether look-down equivalence for BPI spaces should imply BPI equivalence. Indeed it is amusing that we have the aforementioned negative results for mappings from a Cantor set into itself but then a positive result for mappings between Cantor sets (Proposition 11.18). One might do better in trying to pass from information about mappings to information about spaces, rather than to stronger information about the mappings, as in the positive results about big bilipschitz pieces in the context of Euclidean spaces.

14.2 Some examples

Let F be a finite set with at least two elements, and form the "symbolic" Cantor set F^∞ as in Section 2.3. Let $d_a(x, y)$ be the metric on F^∞ defined in (2.4). We do not care what value of a is used here, so long as $0 < a < 1$. For our purposes all of these values of a are equivalent, because of the snowflake functor.

The elements of F^∞ are sequences with values in F, and it will helpful to be specific that they are sequences with index set $\mathbf{Z}_+$.

Let $Sym(\mathbf{Z}_+)$ denote the infinite symmetric group on $\mathbf{Z}_+$, i.e., the group of all bijections from $\mathbf{Z}_+$ to itself. Given $\pi \in Sym(\mathbf{Z}_+)$, define $S_\pi : F^\infty \to F^\infty$ by

$$S_\pi(x) = \{x_{\pi^{-1}(l)}\}_{l=1}^\infty, \quad x = \{x_l\}_{l=1}^\infty. \tag{14.1}$$

One can check that

$$S_\pi \circ S_\sigma = S_{\pi\sigma} \tag{14.2}$$

for all $\pi, \sigma \in Sym(\mathbf{Z}_+)$.

We shall be interested in Lipschitz mappings on F^∞ which arise from permutations in this manner.

Definition 14.2 (Backwardness) *Given $\pi \in Sym(\mathbf{Z}_+)$, its* backwardness $b(\pi)$ *is defined to be the smallest integer k such that*

$$\pi(l) \geq l - k \tag{14.3}$$

for all $l \in \mathbf{Z}_+$. If no such k exists then we set $b(\pi) = \infty$.

Lemma 14.3 *If $\pi \in Sym(\mathbf{Z}_+)$ has finite backwardness, then $S_\pi : F^\infty \to F^\infty$ is Lipschitz with respect to $d_a(x, y)$ with norm $a^{-b(\pi)}$.*

This is easy to check from the definition (2.4) of $d_a(x, y)$.

Lemma 14.4 *$S_\pi : F^\infty \to F^\infty$ is a measure-preserving homeomorphism on F^∞ for all $\pi \in Sym(\mathbf{Z}_+)$. For this we use the measure μ on F^∞ which is the infinite product of uniformly distributed probability measures on each of the individual copies of F.*

This is also easy to check. (For the measure-preserving property one reduces to the case of "cylindrical" sets in which all but finitely many coordinates are free.)

Thus permutations on $\mathbf{Z}_+$ with bounded backwardness induce homeomorphisms on F^∞ which preserve measure and have bounded Lipschitz norm, but without any control on bilipschitzness. For instance, consider the permutation defined by

$$\begin{aligned} \pi(i) &= i - 1 && \text{when } 2 \leq i \leq n \\ &= n && \text{when } i = 1 \\ &= i && \text{when } i > n + 1 \end{aligned} \tag{14.4}$$

This permutation has backwardness $= 1$. However, by having $\pi(1) = n$ where n is as large as we like it can take points that are not close to each other to points that are very close together.

In order to make a better example of this type we put copies of permutations like these on top of themselves. Let $\{(i_p, j_p)\}_{p=1}^\infty$ be a sequence of ordered pairs of positive integers with the properties that $i_p < j_p$ and $i_{p+1} = j_p + 1$ for all p, and also

$$\lim_{p\to\infty} (j_p - i_p) = \infty. \tag{14.5}$$

Define $\pi \in Sym(\mathbf{Z}_+)$ by

$$\begin{aligned} \pi(l) &= l && \text{when } l < i_1 \\ &= j_p && \text{when } l = i_p \\ &= l - 1 && \text{when } i_p < l \le j_p \end{aligned} \tag{14.6}$$

Note that the assumptions on $\{(i_p, j_p)\}_{p=1}^{\infty}$ ensure that $\mathbf{Z}_+$ is the disjoint union of its intersection with $[1, i_1)$ and the intervals $[i_p, j_p]$, $p = 1, 2, 3, \ldots$. This implies that the definition of π just given is complete and consistent.

These choices of $\{(i_p, j_p)\}_{p=1}^{\infty}$ and π should be considered as fixed for the rest of the section.

Proposition 14.5 *If π is as above, then $S_\pi : F^\infty \to F^\infty$ is a homeomorphism which is Lipschitz with norm a^{-1}, measure-preserving, and regular, but does not have a big bilipschitz piece.*

The homeomorphism, Lipschitz, and measure-preserving properties follow from the lemmas above. We conclude regularity from Lemma 12.23, since our measure μ on F^∞ is Ahlfors regular. (This last was mentioned in Section 2.3.)

Thus it remains to show that S_π has no big bilipschitz piece. Suppose to the contrary that E is a closed subset of F^∞ with positive measure on which S_π is bilipschitz.

Let us set some notation. Let m denote the number of elements of F. Given $x \in F^\infty$ and a positive integer k, set

$$\begin{aligned} N_k(x) &= \{y \in F^\infty : d_a(x, y) \le a^k\} \\ &= \{y \in F^\infty : y_l = x_l \text{ when } 1 \le l \le k\} \end{aligned} \tag{14.7}$$

and

$$\begin{aligned} A_k(x) &= N_{k-1}(x) \backslash N_k(x) \\ &= \{y \in F^\infty : d_a(x, y) = a^{k-1}\} \\ &= \{y \in F^\infty : y_l = x_l \text{ when } 1 \le l < k \text{ and } y_k \ne x_k\}. \end{aligned} \tag{14.8}$$

Thus $N_k(x)$ is a small neighborhood of x when k is large, while the "annular" set $A_k(x)$ is disjoint from $N_k(x)$ but still close to it. Notice that $\mu(N_k(x)) = m^{-k}$ and $\mu(A_k(x)) = (m - 1) \cdot m^{-k}$.

Lemma 14.6 *Let $p \in \mathbf{Z}_+$ and $x \in F^\infty$ be given. Then there is a measure-preserving homeomorphism h on F^∞ such that $h(N_{i_p}(x)) \subseteq A_{i_p}(x)$ and*

$$d_a(S_\pi(y), S_\pi(h(y))) \le a^{j_p - i_p}\, d_a(y, h(y)) \tag{14.9}$$

for all $y \in N_{i_p}(x)$.

Thus h helps us to describe the failure of bilipschitzness for S_π.

Let p and x be given as above. Set $u = x_{i_p}$, an element of F, and let $\phi : F \to F$ be a permutation on F such that $\phi(u) \neq u$. Define $h : F^\infty \to F^\infty$ by

$$h(z) = \{\zeta_l\}_{l=1}^{\infty}, \qquad \text{where } \zeta_l = z_l \text{ when } l \neq i_p, \zeta_{i_p} = \phi(z_{i_p}). \tag{14.10}$$

It is easy to see that this is a measure-preserving homeomorphism on F^∞. (It is even an isometry.) It is also easy to verify $h(N_{i_p}(x)) \subseteq A_{i_p}(x)$ from the definitions. We also have that

$$d_a(y, h(y)) = a^{i_p - 1} \qquad \text{when } y \in N_{i_p}(x). \tag{14.11}$$

This follows from the definition (2.4) of $d_a(x, y)$.

Because h disturbs only the i_p^{th} component of its argument, the definition of π ensures that $S_\pi(y)$ and $S_\pi(h(y))$ can differ only in the j_p^{th} component. This means that

$$d_a(S_\pi(y), S_\pi(h(y))) \leq a^{j_p - 1} \tag{14.12}$$

for all $y \in F^\infty$. Thus we get (14.9). This completes the proof of the lemma.

Let us now derive Proposition 14.5 from the lemma. Remember that we are assuming that E is a closed subset of F^∞ with positive measure on which S_π is bilipschitz, and we want to get a contradiction.

Since E has positive measure we can find a point of density of it. This means that there is a point $x \in F^\infty$ such that

$$\lim_{k\to\infty} \frac{\mu(E \backslash N_k(x))}{\mu(N_k(x))} = 0. \tag{14.13}$$

This implies that

$$\lim_{p\to\infty} \frac{\mu(E \backslash N_{i_p}(x))}{\mu(N_{i_p}(x))} = 0 \quad \text{and} \quad \lim_{p\to\infty} \frac{\mu(E \backslash A_{i_p}(x))}{\mu(A_{i_p}(x))} = 0. \tag{14.14}$$

If p is large enough, then we can find $y \in N_{i_p}(x)$ such that $y \in E$ and $h(y) \in E$, where h is as in Lemma 14.6 for this choice of p and x. We are using the fact that h maps $N_{i_p}(x)$ into $A_{i_p}(x)$ and is measure-preserving. On the other hand (14.9) violates the assumption that S_π is bilipschitz on E if p is chosen large enough.

Thus we get a contradiction, and the proposition follows.

Remark 14.7 Although this mapping S_π does not have a big bilipschitz piece, it does have bilipschitz weak tangents. This is not very difficult to check, from the definitions. We shall give a more complicated example later which does not even have any bilipschitz weak tangents.

14.3 Possibilities for Lipschitz mappings

In this section we provide a few simple observations that clarify the possibility of some vestige of bilipschitz behavior in a Lipschitz mapping. There are two main points. The first is to look at what happens if a Lipschitz mapping has no weak tangent which is bilipschitz (since we could often be happy just to have one). The second point is to consider what happens in the presence of some self-similarity assumptions on the Lipschitz mapping.

We want to analyze the failure of bilipschitzness for a mapping. Suppose that we are given metric spaces $(M, d_M(\cdot,\cdot))$, $(N, d_N(\cdot,\cdot))$ and a Lipschitz mapping $f : M \to N$. Let $\epsilon > 0$ and $k > 1$ be given. (Think of ϵ as being small and k as being large.) Define $\mathcal{U}(\epsilon, k)$ to be the set of ordered pairs (x, t) in $M \times (0, \operatorname{diam} M)$ such that

$$\begin{aligned}&\text{there exist } y, z \in \overline{B}_M(x, kt) \text{ with } d_M(y,z) \geq t/k \\ &\text{but } d_N(f(y), f(z)) \leq \epsilon\, d_M(y,z).\end{aligned} \tag{14.15}$$

Roughly speaking, $\mathcal{U}(\epsilon, k)$ represents the set of locations and scales in M at which there is substantial collapsing of points by f, where the amount of collapsing is measured by ϵ, and where k reflects the degree of resolution in the notion of "location and scale".

Definition 14.8 (Thoroughly unbilipschitz mappings (TUMs)) *With the notations and assumptions described above, we say that f is a* TUM *("thoroughly unbilipschitz mapping") if for each $\epsilon > 0$ there exists $k = k(\epsilon)$ so that $\mathcal{U}(\epsilon, k)$ is all of $M \times (0, \operatorname{diam} M)$.*

Examples 14.9 (a) Constant mappings are TUMs.

(b) The mapping $f : \mathbf{R}^2 \to \mathbf{R}$ given by $f(x) = x_1$ is a TUM.

(c) If M and N are Ahlfors regular and the dimension of M is larger than the dimension of N, then any Lipschitz mapping from M to N is a TUM. (Exercise.)

(d) The identity as a mapping from $([0,1], |x-y|^\alpha)$ to $([0,1], |x-y|)$ is a TUM when $\alpha < 1$.

(e) The mappings considered in Sections 11.3 and 11.6 are more interesting examples of TUMs.

This definition is somewhat technical, but it has the redeeming feature that it admits the following characterization.

Lemma 14.10 *Let $(M, d_M(\cdot,\cdot))$ and $(N, d_N(\cdot,\cdot))$ be two metric spaces which are doubling and complete, and assume that $f : M \to N$ is Lipschitz. Assume also that M is uniformly perfect (Definition 5.3). Then f is a TUM if and only if none of the weak tangents to f is bilipschitz.*

The assumption of uniform perfectness prevents M from having islands that are too isolated. This would be a nuisance for the purpose of the lemma.

The proof of the lemma is straightforward but tedious in complete detail and so we only sketch the main points. It is not very difficult to show that if f is a

TUM then for each $\epsilon > 0$ and for each weak tangent $g : M' \to N'$ of f there exist $y', z' \in M'$ with $y' \neq z'$ such that

$$d_{N'}(g(y'), g(z')) \leq \epsilon \, d_{M'}(y', z'). \tag{14.16}$$

Indeed, by definition g is obtained as the limit of a sequence of renormalizations of f to various locations and scales, and the TUM condition implies the existence of a pair of points with severe collapsing near all locations and scales.

Conversely, if f is not TUM, then there is an $\epsilon > 0$ so that for each $k = 1, 2, 3, \ldots$ one can find $(x_k, t_k) \notin \mathcal{U}(\epsilon, k)$. We want to blow up f along this sequence. That is, we now consider f as a mapping from M_{x_k,t_k} to N_{ξ_k,t_k}, where $\xi_k = f(x_k)$, and where the pointed metric spaces M_{x_k,t_k} and N_{ξ_k,t_k} are defined as in Section 9.1. By passing to a subsequence we can conclude that this sequence of mappings converges to a weak tangent of f, as in Section 9.7, and one can check that the weak tangent must be bilipschitz. There is one minor technical problem however. According to our definition of weak tangents we should also require that $t_k \leq \operatorname{diam} N$ for all k. This is only an issue when $\operatorname{diam} N$ is finite. If the t_k's remain bounded, then it is easy enough to make a modest renormalization to get $t_k \leq \operatorname{diam} N$ with the only change that now $(x_k, t_k) \notin \mathcal{U}(\epsilon, L^{-1} k)$ for some constant L. This suffices for the conclusion that the weak tangent be bilipschitz. If the t_k's do not remain bounded, then $\operatorname{diam} M = \infty$, and for each k there is a point $y_k \in M$ such that $C^{-1} t_k \leq d_M(x_k, y_k) \leq t_k$, where C does not depend on k. This follows from the assumption that M is uniformly perfect. The approximate bilipschitz condition implied by $(x_k, t_k) \notin \mathcal{U}(\epsilon, k)$ then forces $\operatorname{diam} N = \infty$ too, so that there is no problem. This completes the proof of the lemma.

Next we define an approximate self-similarity condition for mappings which is analogous to the BPI condition for spaces.

Definition 14.11 (BPI for mappings) *Let $(M, d_M(\cdot,\cdot))$ and $(N, d_N(\cdot,\cdot))$ be metric spaces, and suppose that $f : M \to N$ is Lipschitz. Assume that M is Ahlfors regular of dimension d. We say that f has* BPI *if there exist $\theta, k > 0$ so that for each pair of balls $B_M(x_1, r_1)$ and $B_M(x_2, r_2)$ in M, with $0 < r_1, r_2 \leq \operatorname{diam} M$, there is a closed set $A \subseteq B_M(x_1, r_1)$ with $H^d(A) \geq \theta \, r_1^d$, and mappings $\phi : A \to B_M(x_2, r_2)$ and $\psi : f(A) \to f(\phi(A))$ such that ϕ and ψ are k-conformally bilipschitz with scale factor r_2/r_1 and $\psi \circ f = f \circ \phi$ on A.*

For this to hold it is necessary that M be BPI, but not N.

Examples 14.12 (a) Any constant mapping on a BPI space is a BPI mapping.

(b) Bilipschitz mappings on a BPI space are BPI mappings.

(c) The mapping $f : \mathbf{R}^2 \to \mathbf{R}$ given by $f(x) = x_1$ is BPI.

(d) The mappings considered in Sections 11.3 and 11.6 are BPI mappings.

(e) If M is BPI and $f : M \to N$ is Lipschitz and has BBP (Definition 7.20), then f has BPI. To prove this one uses Propositions 7.21 and 6.9.

Lemma 14.13 (Weak tangents of BPI mappings) *Let M and N be metric spaces, with M Ahlfors regular of dimension d, and assume that $f : M \to N$ is Lipschitz and BPI. Then any weak tangent $g : M' \to N'$ to f also has BPI. In fact, f and g have "big pieces of each other", which means that there exist $\theta, k > 0$ so that for each pair of balls $B_M(x,r)$ and $B_{M'}(x',r')$ in M and M', respectively, with $0 < r \leq \operatorname{diam} M$ and $0 < r' \leq \operatorname{diam} M'$, there is a closed set $A \subseteq B_M(x,r)$ with $H^d(A) \geq \theta\, r_1^d$, and mappings $\phi : A \to B_{M'}(x',r')$ and $\psi : f(A) \to g(\phi(A))$, such that ϕ and ψ are k-conformally bilipschitz with scale factor r_2/r_1 and $\psi \circ f = g \circ \phi$ on A.*

The proof is messy but straightforward, using compactness. It is also very similar to the arguments used for Proposition 9.8 and Corollary 9.9, and we omit the details.

Proposition 14.14 *If $f : M \to N$ has BPI, then it either has BBP or it is a TUM.*

Indeed, either f has a bilipschitz weak tangent, or it does not. If not, then f is a TUM, by Lemma 14.10. If f does have a bilipschitz weak tangent $g : M' \to N'$, then f must have BBP, because f and g have big pieces of each other, as in Lemma 14.13.

14.4 A stronger example

We saw in Section 14.2 how there exist homeomorphisms on Cantor sets which are Lipschitz, measure preserving, and regular, but which do not have a big bilipschitz piece. These examples did have some bilipschitz weak tangents, however. In this section we construct a mapping which is a TUM and BPI. In fact it enjoys stronger self-similarity properties than BPI, and has double points at all locations and scales.

For this example we shall use the symbolic Cantor set F^∞ with $F = \{0, 1\}$. For the sake of definiteness let us use the metric $d(x, y)$ on F^∞ which comes from (2.4) with $a = 1/2$. As before we use the measure μ on F^∞ which is the infinite product of the uniform "coin tosses" on the various copies of F. Let $N_k(x)$ and $A_k(x)$ be as defined in (14.7) and (14.8) (with $a = 1/2$).

Let $a = \{a_j\}_{j=1}^\infty$ and $b = \{b_j\}_{j=1}^\infty$ denote the elements of F^∞ which are defined by $a_j = 0$ for all j and $b_1 = 1$, $b_j = 0$ for all $j \geq 2$.

Define $\beta : F^\infty \to F^\infty$ as follows. First we set

$$\beta(a) = \beta(b) = a. \tag{14.17}$$

For each $k \geq 2$ we map $A_k(a) \cup A_k(b)$ onto $A_{k-1}(a)$ in the following manner. By definition,

$$A_{k-1}(a) = \{x \in F^\infty : x_j = 0 \text{ when } j < k-1 \text{ and } x_{k-1} = 1\}. \tag{14.18}$$

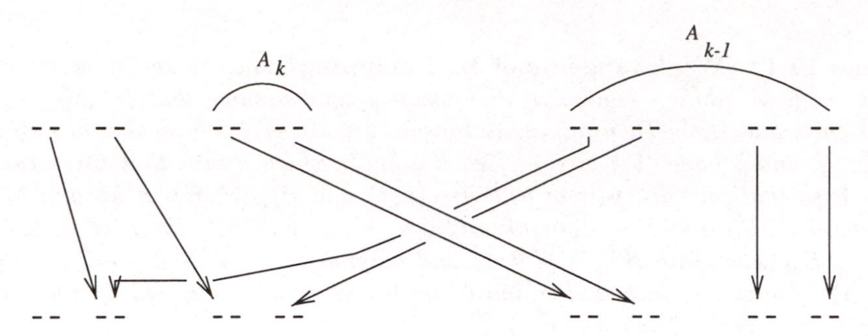

FIG. 14.1. A diagram for the definition of β

Define $A^i_{k-1}(a)$ for $i = 0, 1$ by

$$A^i_{k-1}(a) = \{x \in F^\infty : x_j = 0 \text{ when } j < k-1, \\ x_{k-1} = 1, \text{ and } x_k = i\}. \tag{14.19}$$

Thus $A_{k-1}(a) = A^0_{k-1}(a) \cup A^1_{k-1}(a)$. We define β so that it maps $A_k(a)$ onto $A^0_{k-1}(a)$ and $A_k(b)$ onto $A^1_{k-1}(a)$ in the obvious manner. Explicitly, if $x \in A_k(a)$, $k \geq 2$, then

$$\begin{aligned} \beta(x)_j &= x_j && \text{when } j \neq k-1, k \\ &= 1 && \text{when } j = k-1 \\ &= 0 && \text{when } j = k \end{aligned} \tag{14.20}$$

where $\beta(x)_j$ denotes the jth component of $\beta(x) \in F^\infty$. If $x \in A_k(b)$, $k \geq 2$, then we set

$$\begin{aligned} \beta(x)_j &= 0 && \text{when } j = 1 \text{ and } j \neq k-1 \\ &= x_j && \text{when } j \neq 1, k-1, k \\ &= 1 && \text{when } j = k-1 \\ &= 1 && \text{when } j = k. \end{aligned} \tag{14.21}$$

Thus the restriction of β to each of $A_k(a)$ and $A_k(b)$ is really just a copy of the identity but with a couple of changes in the beginning coordinates (which ensure that these sets are mapped to the correct places).

This completes the definition of β, because

$$F^\infty = \{a, b\} \cup \Big(\bigcup_{k=2}^{\infty} (A_k(a) \cup A_k(b)) \Big). \tag{14.22}$$

Note that these unions are disjoint, so that β is well defined. Of course the $A_{k-1}(a)$'s are also pairwise disjoint, and we have that

$$F^\infty = \{a\} \cup \bigcup_{k=2}^{\infty} A_{k-1}(a). \tag{14.23}$$

Lemma 14.15 *$\beta : F^\infty \to F^\infty$ is surjective, injective on $F^\infty \setminus \{a, b\}$, and*

$$\beta(A_k(a)) = A^0_{k-1}(a), \qquad \beta(A_k(b)) = A^1_{k-1}(a),$$
$$\beta^{-1}(A^0_{k-1}(a)) = A_k(a), \qquad \beta^{-1}(A^1_{k-1}(a)) = A_k(b)$$

for each $k \geq 2$. Moreover, β is 2-Lipschitz, measure-preserving, and an isometry on each $A_k(a)$ and $A_k(b)$, $k \geq 2$.

This is not hard to check from the definitions. The "2" in "2-Lipschitz" comes from the introduction of the 1 in the $j = k - 1$ coordinate of $\beta(x)$ in (14.20) and (14.21) above. Thus $d(\beta(x), \beta(a)) = 2\, d(x, a)$ when $x \in A_k(a)$, for instance.

β has a single double point, $\beta(a) = \beta(b) = a$. We want to make a mapping with double points at all scales and locations, and so we iterate the construction.

Let us call a set $I \subseteq F^\infty$ a *cell* if it is of the form $N_k(x)$ for some $x \in F^\infty$ and $k \geq 0$. We shall sometimes say *k-cell* to be more explicit. Notice that any pair of cells is either disjoint or one is contained in the other.

For each cell $I \subseteq F^\infty$ we define a mapping $\beta_I : I \to I$ exactly as above. That is, there is a canonical bijection ψ_I from the k-cell I onto F^∞, namely $\{y_j\} \mapsto \{y_{j+k}\}$, and we use ψ_I to transport β back to I. Note that $\beta_I(I) = I$ and that β_I is 2-Lipschitz.

We want to define a sequence of mappings $f_n : F^\infty \to F^\infty$ obtained by putting copies of β on top of themselves. To do this we first define some auxiliary mappings and some families of cells. We start with

$$g_0 = \beta \tag{14.24}$$

and

$$\mathcal{F}_1 = \{\text{the cells } A^0_{k-1} \text{ and } A^1_{k-1}, k \geq 2\} \tag{14.25}$$

Notice that the elements of $\mathcal{F}_1$ are pairwise disjoint and that their union is $F^\infty \setminus \{a\}$.

Next we define $g_1 : F^\infty \to F^\infty$ by

$$\begin{aligned} g_1 &= \beta_I \qquad \text{on } I \text{ when } I \in \mathcal{F}_1, \\ g_1(a) &= a. \end{aligned} \tag{14.26}$$

In general we shall define for each $j \geq 1$ a countable family of cells $\mathcal{F}_j$, in such a way that the elements of $\mathcal{F}_j$ are pairwise disjoint and their union is dense in F^∞. Once $\mathcal{F}_j$ is specified we define $g_j : F^\infty \to F^\infty$ by

$$\begin{aligned} g_j &= \beta_I \qquad \text{on } I \text{ for each } I \in \mathcal{F}_j, \\ g_j(x) &= x \qquad \text{when } x \in F^\infty \setminus \bigcup_{I \in \mathcal{F}_j} I. \end{aligned} \tag{14.27}$$

We define $\mathcal{F}_j$ recursively by the rule

$$\mathcal{F}_{j+1} = \bigcup_{I \in \mathcal{F}_j} \mathcal{F}_1(I), \tag{14.28}$$

where $\mathcal{F}_1(I)$ is the pull-back of $\mathcal{F}_1$ from F^∞ to I using ψ_I. Thus the elements of $\mathcal{F}_{j+1}$ are pairwise disjoint,

$$F^\infty \setminus \bigcup_{I \in \mathcal{F}_{j+1}} I \qquad \text{is countable} \tag{14.29}$$

(by induction, since this set $= \{a\}$ when $j = 0$), and their union is dense in F^∞.

In this manner we can define $\mathcal{F}_j$ and g_j for all j.

Given a k-cell $I \subseteq F^\infty$, let us call a mapping $h : I \to F^\infty$ *rigid* if $h(I)$ is also a k-cell, and if $h(x)$ is defined for $x \in I$ in such a way that $h(x)_j = x_j$ when $j > k$, $h(x)_j = \eta_j$ when $1 \leq j \leq k$, where η_j does not depend on x, only on h. Thus rigidity means that h basically looks like the identity, modulo some sliding around. If h is a mapping on all of F^∞, then we say that h is rigid on a cell if its restriction to the cell is a rigid mapping, and we call a cell a *rigid cell for h* in this case. By construction $\mathcal{F}_1$ consists of *images* under β of rigid cells for β. More generally $\mathcal{F}_1(I)$ consists of cells which are the images of rigid cells for β_I. Similarly the cells in $\mathcal{F}_{j+1}$ are the images under g_j of cells which are rigid for g_j.

Next we define $f_j : F^\infty \to F^\infty$ for $j \geq 0$ by

$$f_j = g_j \circ g_{j-1} \circ \cdots \circ g_0. \tag{14.30}$$

In the next lemmas we try to bring out some combinatorial properties of these mappings. The main point is that the actions of the g_j's are nested in a good way, and this will permit us to control f_j even for very large j.

Lemma 14.16 *For each $I \in \mathcal{F}_j$ and $l < j$ there is a unique $L \in \mathcal{F}_l$ such that $L \supseteq I$, $L \neq I$. I is disjoint from all other elements of $\mathcal{F}_l$.*

This follows from the definitions. The main point is that each element of $\mathcal{F}_1(I)$ is a proper subset of I. The last part follows because the elements of $\mathcal{F}_l$ are pairwise disjoint.

Lemma 14.17 *If $1 \leq l \leq k$ and $L \in \mathcal{F}_l$, then $g_k(L) = L$ and $g_k^{-1}(L) = L$.*

It is clear from the definition that $g_l(L) = L$ when $L \in \mathcal{F}_l$. This comes from the fact that $\beta_I(I) = I$, which itself follows from the surjectivity of β. We also get that $g_l^{-1}(L) = L$ from the definition, because the elements of $\mathcal{F}_l$ are pairwise disjoint, and so they cannot interact with each other. When $k > l$ the lemma follows from the $k = l$ case, the nesting properties provided by Lemma 14.16, and the fact that g_k equals the identity off the elements of $\mathcal{F}_k$.

Lemma 14.18 *For each $j \geq 0$, $g_j^{-1}(J)$ is a cell for all $J \in \mathcal{F}_{j+1}$, g_j is rigid on this cell, and $g_j(g_j^{-1}(J)) = J$.*

When $j = 0$ this is a statement about β which follows from the definitions. See Lemma 14.15. The general case follows from the corresponding statement for β, because of the way that g_j is defined. Lemma 14.17 is also relevant, to ensure that the various pieces of g_j on the elements of $\mathcal{F}_j$ do not interact.

Lemma 14.19 *For each $j \geq 1$ and $J \in \mathcal{F}_{j+1}$ there is a $K \in \mathcal{F}_j$ such that $g_j^{-1}(J) \subseteq K$.*

This follows easily from the construction.

Lemma 14.20 *For each $l \geq 0$ and $L \in \mathcal{F}_{l+1}$ we have that $f_l^{-1}(L)$ is a cell, f_l is rigid on that cell, and $f_l(f_l^{-1}(L)) = L$.*

This follows from the previous lemmas. That is, we repeat the corresponding facts for the g_j's, we use the nesting properties for these cells, and we use the fact that if a mapping ϕ is rigid on a cell I, and if we take any cell J contained in $\phi(I)$, then $\phi^{-1}(J)$ is a cell on which ϕ is rigid. We also use the fact that if ψ is a rigid mapping on J, then $\psi \circ \phi$ is rigid on $\phi^{-1}(J)$.

Lemma 14.21 *If $j \geq 1$ and $I \in \mathcal{F}_{j+1}$, then $f_m(f_j^{-1}(I)) = I$ when $m \geq j$.*

Indeed, Lemma 14.20 implies that $f_j(f_j^{-1}(I)) = I$ ("$\subseteq$" is automatic, the lemma is used to get "$\supseteq$"). For $m > j$ one reduces to this observation using Lemma 14.17.

Lemma 14.22 *Each f_j is 2-Lipschitz.*

To prove this we fix arbitrary choices of $j \geq 0$ and $x, y \in F^\infty$. For each $l \geq 0$ we set

$$\begin{aligned} R_l(x,y) &= 1 \qquad \text{if there is an } I \in \mathcal{F}_{l+1} \text{ with } f_l(x), f_l(y) \in I \\ &= 0 \qquad \text{otherwise.} \end{aligned} \tag{14.31}$$

Lemma 14.20 implies that

$$d(f_l(x), f_l(y)) = d(x,y) \qquad \text{when} \qquad R_l(x,y) = 1. \tag{14.32}$$

In particular we are finished if $R_j(x,y) = 1$, and we assume that $R_j(x,y) = 0$.

We shall need the fact that

$$R_k(x,y) = 1 \qquad \text{when} \qquad R_l(x,y) = 1 \text{ and } k \leq l. \tag{14.33}$$

This is an easy consequence of Lemma 14.19.

Suppose now that $R_l(x,y) = 1$ for some l, and let us take l as large as possible. The preceding observation ensures that the maximal l is less than j. We have that $f_l(x), f_l(y) \in I$ for some $I \in \mathcal{F}_{l+1}$, and we know that $d(f_l(x), f_l(y)) = d(x,y)$.

To understand the behavior of $f_i(x), f_i(y)$ for $i > l$ we distinguish three cases.

Case 1.

If $f_{l+1}(x), f_{l+1}(y) \in F^\infty \backslash \bigcup_{J \in \mathcal{F}_{l+2}} J$, then $f_i(x) = f_{l+1}(x)$ and $f_i(y) = f_{l+1}(y)$ for all $i \geq l+1$, and hence

$$d(f_j(x), f_j(y)) = d(f_{l+1}(x), f_{l+1}(y)). \tag{14.34}$$

On the other hand, the restriction of g_{l+1} to I is 2-Lipschitz, because g_{l+1} equals β_I on I, and this is essentially a carbon copy of β. Thus

$$\begin{aligned} d(f_{l+1}(x), f_{l+1}(y)) &= d(g_{l+1}(f_l(x)), g_{l+1}(f_l(y))) \\ &\leq 2\, d(f_l(x), f_l(y)) = 2\, d(x, y). \end{aligned} \tag{14.35}$$

Thus we get the required bound for $d(f_j(x), f_j(y))$ in this case.

Before we proceed with the argument let us record a simple fact.

Lemma 14.23 *If K_1 and K_2 are disjoint cells in F^∞, then $d(u_1, u_2)$ for $u_i \in K_i$ depends only on K_1 and K_2 and not on the individual choices of u_1 or u_2. If $v \in F^\infty \backslash K_1$, then $d(v, u_1)$ is independent of the choice of $u_1 \in K_1$.*

This is easy to verify from the definitions of cells and the metric $d(\cdot,\cdot)$.

Case 2.

Suppose now that $f_{l+1}(x)$ and $f_{l+1}(y)$ both lie in $\bigcup_{J \in \mathcal{F}_{l+2}} J$. They must lie in different cells $J_x, J_y \in \mathcal{F}_{l+2}$, since otherwise l was not maximal (i.e., we would have $R_{l+1}(x,y) = 1$). Lemma 14.21 yields $f_i(x) \in J_x$ and $f_i(y) \in J_y$ for all $i \geq l+1$. Lemma 14.23 implies now that (14.34) holds, since J_x and J_y must be disjoint. Once we have this we can finish as before.

Case 3.

We are left with the case where exactly one of $f_{l+1}(x)$ and $f_{l+1}(y)$ lies in $\bigcup_{J \in \mathcal{F}_{l+2}} J$. Without loss of generality we assume that $f_{l+1}(x)$ belongs to some $J_x \in \mathcal{F}_{l+2}$ while $f_{l+1}(y)$ does not belong to any such cell. Then $f_j(y) = f_l(y)$, while $f_j(x) \in J_x$, for the same reasons as before. This permits us to conclude again that (14.34) holds, and the rest of the argument is the same as before.

This finishes the analysis of the three cases, which cover all possibilities when $R_l(x,y) = 1$ for some l. Thus we may assume now that $R_0(x,y) = 0$. In this situation $f_0(x)$ and $f_0(y)$ cannot both lie in a single element of $\mathcal{F}_1$. Either they both lie in $F^\infty \backslash \bigcup_{J \in \mathcal{F}_1} J$, or neither of them does (and they lie in distinct cells in $\mathcal{F}_1$), or exactly one of them does. In each case the argument is the same as before. The main point is to show that

$$d(f_j(x), f_j(y)) = d(f_0(x), f_0(y)). \tag{14.36}$$

and then to reduce to the fact that $f_0 = g_0 = \beta$ is 2-Lipschitz.

This completes the proof of Lemma 14.22.

Lemma 14.24 *Each $f_j : F^\infty \to F^\infty$ is one-to-one except on a countable set, surjective, and measure-preserving.*

To prove this it suffices to show that the analogous statements hold for each g_j. For g_j they follow from the definition and the analogous statements for β.

Lemma 14.25 $d(f_l(x), f_{l+1}(x)) \leq 4^{-l-1}$ *for all* $l \geq 0$ *and* $x \in F^\infty$.

To see this it suffices to show that

$$d(g_{l+1}(y), y) \leq 4^{-l-1} \tag{14.37}$$

for all $l \geq 0$ and $y \in F^\infty$. Because of the definition of g_j, the left side of this inequality is nonzero only when $y \in I$ for some $I \in \mathcal{F}_{l+1}$, and in this case we have $g_{l+1}(y) \in I$ too. Thus we are reduced to the statement that

$$\operatorname{diam} I \leq 4^{-j} \qquad \text{when } I \in \mathcal{F}_j, j \geq 1. \tag{14.38}$$

For $\mathcal{F}_1$ this is true by inspection of the definition. The general case follows from this one and the recursive definition of $\mathcal{F}_j$.

This lemma implies that the f_j's converge uniformly to a mapping $f : F^\infty \to F^\infty$. More generally, define mappings $h_{l,m} : F^\infty \to F^\infty$ by

$$h_{l,m} = g_m \circ g_{m-1} \circ \cdots \circ g_{l+1}. \tag{14.39}$$

The $h_{l,m}$'s converge uniformly as $m \to \infty$ to a mapping $h_l : F^\infty \to F^\infty$.

We need to show that f and the h_l's behave like the f_m's and the $h_{l,m}$'s, i.e., that we do not lose the nice properties in taking the limit.

Lemma 14.26 *f and the h_l's are 2-Lipschitz.*

For f this follows from Lemma 14.22. The analogue of Lemma 14.22 also holds for the h's, i.e., each $h_{l,m}$ is also 2-Lipschitz. The proof is essentially the same, only cosmetic changes to the argument are needed. Thus the h_l's are 2-Lipschitz too.

Lemma 14.27 $h_{l,m}(I) = I$ *for all* $I \in \mathcal{F}_{l+1}$.

This follows easily from Lemma 14.17.

Lemma 14.28 $h_l(I) = I$ *for all* $I \in \mathcal{F}_{l+1}$, *and* $h_l(x) = x$ *when* x *lies in* $F^\infty \setminus \bigcup_{I \in \mathcal{F}_{l+1}} I$.

The first part follows from the preceding lemma and compactness. That is, $h_l(I) \subseteq I$ follows immediately from the previous lemma, but to get $h_l(I) = I$ one uses the compactness of I and the uniform convergence of the $h_{l,m}$'s as $m \to \infty$. For the second part it suffices to show that

$$h_{l,m}(x) = x \qquad \text{when } x \in F^\infty \setminus \bigcup_{I \in \mathcal{F}_{l+1}} I \tag{14.40}$$

when $m > l$. This in turn follows from the corresponding property of the g_j's and the observation that $\bigcup_{J \in \mathcal{F}_j} J$ decreases as j increases.

Lemma 14.29 $f(f_l^{-1}(I)) = I$ *when* $l \geq 0$ *and* $I \in \mathcal{F}_{l+1}$.

This can be derived from Lemma 14.21. (Again, one can use uniform convergence and compactness to get $f(f_l^{-1}(I)) \supseteq I$.)

Our next goal is to establish the measure-preserving properties of f.

Lemma 14.30 *If* L *is a* k*-cell in* F^∞, *then* $\operatorname{diam} L = 2^{-k}$ *and (for* $k > 0$*)* $\operatorname{dist}(L, F^\infty \backslash L) = 2^{-k+1}$.

This follows from the definitions.

Lemma 14.31 *If* L *is a* k*-cell in* F^∞, *then* L *can be realized as a union of sets of the form* $f_k^{-1}(I)$, $I \in \mathcal{F}_{k+1}$, *together with a subset of the set* $F^\infty \backslash \bigcup_{I \in \mathcal{F}_{k+1}} f_k^{-1}(I)$. *This last set is itself countable.*

To see this we observe that

$$\operatorname{diam} f_k^{-1}(I) = \operatorname{diam} I \leq 4^{-k-1} \qquad \text{when } I \in \mathcal{F}_{k+1}. \tag{14.41}$$

This follows from Lemma 14.20 (for the equality) and (14.38). Combining this with the preceding lemma we obtain that if L is a k-cell, then for each $I \in \mathcal{F}_{k+1}$ we have either that $f_k^{-1}(I)$ is contained in L or it is disjoint from L. This implies that I can be realized as a union in the manner described in the lemma. The countability assertion follows from (14.29) and the fact that f_k is injective except on a countable set (Lemma 14.24). This proves the lemma.

Lemma 14.32 *If* L *is a* k*-cell in* F^∞, *then* $f(L) = f_j(L)$ *for all* $j \geq k$.

Indeed, given L we can write it as a union in the manner of Lemma 14.31. If $x \in F^\infty \backslash \bigcup_{I \in \mathcal{F}_{k+1}} f_k^{-1}(I)$, then $f_j(x) = f_k(x)$ for all $j \geq k$ (because $g_m(y) = y$ when $y \in F^\infty \backslash \bigcup_{I \in \mathcal{F}_{k+1}} I$ and $m \geq k+1$), and hence $f(x) = f_k(x)$ for these x's. Thus these points cause no trouble for our assertion that $f(L) = f_j(L)$. If $I \in \mathcal{F}_{k+1}$, then $f(f_k^{-1}(I)) = I = f_j(f_k^{-1}(I))$, because of Lemmas 14.21 and 14.29. Lemma 14.32 follows easily from this observations and Lemma 14.31.

Lemma 14.33 *If* L *is a cell in* F^∞, *then* $\mu(f(L)) = \mu(L)$.

This follows from Lemmas 14.32 and 14.24.

Lemma 14.34 *If* K *and* L *are disjoint cells in* F^∞, *then* $f(K) \cap f(L)$ *is at most countable.*

This also follows from Lemmas 14.32 and 14.24.

Lemma 14.35 *If* U *is an open subset of* F^∞, *then* $f(U)$ *is* σ*-compact (and therefore measurable), and* $\mu(f(U)) = \mu(U)$.

Indeed, any open subset U of F^∞ can be realized as the countable *disjoint* union of cells, namely the maximal cells contained in U. (Given any pair of cells, either one is contained in the other, or they are disjoint.) Thus $f(U)$ is σ-compact, and its measure can be computed using the preceding pair of lemmas. (Of course countable sets have measure zero.)

Lemma 14.36 *$\mu(f(A)) = \mu(A)$ for any compact subset A of F^∞.*

Let $\{U_j\}$ be any *decreasing* sequence of open subsets of F^∞ such that $A = \bigcap_j U_j$. Then

$$\lim_{j\to\infty} \mu(U_j) = \mu(A) \tag{14.42}$$

by a standard exercise in measure theory. Also,

$$\begin{aligned} 0 \le \mu(f(U_j)) - \mu(f(A)) &= \mu(f(U_j)\backslash f(A)) \\ &\le \mu(f(U_j\backslash A)) = \mu(U_j\backslash A) \end{aligned} \tag{14.43}$$

(since $U_j\backslash A$ is open), and the latter quantity tends to 0 as $j \to \infty$. Thus $\lim_{j\to\infty} \mu(f(U_j)) = \mu(f(A))$, and Lemma 14.36 follows from Lemma 14.35.

Let us now "complete" the σ-algebra of Borel sets in F^∞ with respect to μ in the standard way, by including subsets of Borel sets of μ-measure 0. Thus we shall say that a subset E of F^∞ is *μ-measurable* if there exist Borel sets $X, Y \subseteq F^\infty$ such that $X \subseteq E \subseteq Y$ and $\mu(Y\backslash X) = 0$. This defines a σ-algebra of subsets of F^∞, and we can extend μ to the μ-measurable sets by putting $\mu(E) = \mu(X) = \mu(Y)$.

Lemma 14.37 *If $E \subseteq F^\infty$ is μ-measurable, then so is $f(E)$, and $\mu(f(E)) = \mu(E)$.*

Indeed, let a μ-measurable set E be given. Because μ is Borel regular, we can find an increasing sequence of compact sets $\{A_j\}$ and a decreasing sequence of open sets $\{U_j\}$ in F^∞ such that $A_j \subseteq E \subseteq U_j$ for all j and

$$\lim_{j\to\infty} \mu_j(U_j) = \mu(E) = \lim_{j\to\infty} \mu_j(A_j). \tag{14.44}$$

Lemmas 14.35 and 14.36 imply that $f(A_j)$ and $f(U_j)$ are Borel sets which satisfy

$$\lim_{j\to\infty} \mu_j(f(U_j)) = \mu(E) = \lim_{j\to\infty} \mu_j(f(A_j)). \tag{14.45}$$

Of course the $f(U_j)$'s decrease to a superset of $f(E)$ and the $f(A_j)$'s increase to a subset of $f(E)$, and it follows that $f(E)$ is measurable and has the same measure as E. This proves the lemma.

This completes our discussion of the measure-preserving property for f. Our next goal is to bring out the behavior of f at various scales and locations, and then the self-similarity properties of f.

Lemma 14.38 *If K is a cell in F^∞, then there is an $l \ge 0$ and an $I \in \mathcal{F}_{l+1}$ such that $K \supseteq f_l^{-1}(I)$ and $\operatorname{diam} f_l^{-1}(I) \ge \frac{1}{4} \operatorname{diam} K$.*

The idea here is that our previous lemmas tell us a lot about the behavior of f on a set like $f_l^{-1}(I)$, $I \in \mathcal{F}_{l+1}$, and we are now saying that we can find such a set inside any cell, and as a large proportion of the cell.

Let us prove the lemma. From Lemma 14.31 we certainly know that there is an $l \geq 0$ and an $I \in \mathcal{F}_{l+1}$ such that $K \supseteq f_l^{-1}(I)$. Choose l and I so that $\operatorname{diam} I$ is as large as possible. We need to show that $\operatorname{diam} f_l^{-1}(I) \geq \frac{1}{4} \operatorname{diam} K$.

Let us call $I' \in \mathcal{F}_{l+1}$ a *sibling* of I if they have the same parent L in $\mathcal{F}_l$. (Remember Lemma 14.16.) If $l = 0$, then we take $L = F^\infty$, and we consider all the elements of $\mathcal{F}_1$ to be siblings of I. We have that

$$K \not\supseteq f_{l-1}^{-1}(L) \qquad \text{if } l > 0, \tag{14.46}$$

since we chose I to be as large as possible. On the other hand,

$$f_l^{-1}(I) \subseteq f_{l-1}^{-1}(L), \tag{14.47}$$

because $g_l^{-1}(I) \subseteq g_l^{-1}(L) = L$, by Lemma 14.17. Thus K intersects $f_{l-1}^{-1}(L)$, since it contains $f_l^{-1}(I)$.

From Lemma 14.20 we have that $f_{l-1}^{-1}(L)$ is a cell. Any pair of cells in F^∞ are either disjoint, or one is contained in the other. We know that K intersects $f_{l-1}^{-1}(L)$ and does not contain it (when $l > 0$), and so we conclude that

$$K \subseteq f_{l-1}^{-1}(L) \qquad \text{when } l > 0. \tag{14.48}$$

When $l = 0$ the natural counterpart of this is the automatic inclusion $K \subseteq F^\infty$.

If the diameter of I is not smaller than the diameter of any of its siblings, then $\operatorname{diam} I \geq \frac{1}{4} \operatorname{diam} L$. This follows from the definition of $\mathcal{F}_1$ (and of $\mathcal{F}_1(L)$). In this case we would get the desired inequality $\operatorname{diam} f_l^{-1}(I) \geq \frac{1}{4} \operatorname{diam} K$ because Lemma 14.20 implies that $f_l^{-1}(I)$ and I have the same diameter, while (14.48) ensures that $\operatorname{diam} K \leq \operatorname{diam} f_{l-1}^{-1}(L) \leq \operatorname{diam} L$.

Suppose instead that I has a sibling I' which satisfies $\operatorname{diam} I' > \operatorname{diam} I$. We may as well choose I' so that $\operatorname{diam} I' = 2 \operatorname{diam} I$ and $\operatorname{dist}(f_l^{-1}(I), f_l^{-1}(I'))$ is as small as possible. Notice that there are exactly two choices of siblings $I' \in \mathcal{F}_{l+1}$ with $\operatorname{diam} I' = 2 \operatorname{diam} I$. One can check this by going back to the definition of $\mathcal{F}_1$. By choosing the one that minimizes this distance we can get that

$$\operatorname{dist}(f_l^{-1}(I), f_l^{-1}(I')) \leq 4 \operatorname{diam} I. \tag{14.49}$$

To see this one should start with the case where $l = 0$, so that f_l is just β. In this case the inequality follows from the construction. If $l > 0$ then we should think of f_l as $g_l \circ f_{l-1}$. We have that $g_l^{-1}(I)$ and $g_l^{-1}(I')$ are both contained in the common parent L of I, I', and f_{l-1} is rigid on the cell $f_{l-1}^{-1}(L)$ by Lemma 14.20. Thus for $l > 0$ we can reduce to the problem of showing that

$$\operatorname{dist}(g_l^{-1}(I), g_l^{-1}(I')) \leq 4 \operatorname{diam} I. \tag{14.50}$$

This can be derived from the $l = 0$ case, because $g_l = \beta_L$ on L, and β_L is just a replica of β on L.

Because we chose I so that it is as large as possible, we have that

$$f_l^{-1}(I') \not\subseteq K. \tag{14.51}$$

We know from Lemma 14.20 that $f_l^{-1}(I')$ is a cell, and so either it contains K or it is disjoint from K. If it contains K, then it contains $f_l^{-1}(I)$, and this is impossible because $I, I' \in \mathcal{F}_{l+1}$ are disjoint. Thus K and $f_l^{-1}(I')$ are disjoint. Hence

$$2 \operatorname{diam} K \leq \operatorname{dist}(K, f_l^{-1}(I')), \tag{14.52}$$

by Lemma 14.30. Since K contains $f_l^{-1}(I)$ this last distance is not greater than the one in (14.49), and so we conclude that $\operatorname{diam} K \leq 2 \operatorname{diam} I$, which is fine.

This completes the proof of the lemma.

Before we proceed let us introduce some more terminology. If I is a k-cell, $I = N_k(x)$ for some $x \in F^\infty$, then we call $y \in N_k(x)$ the *left end-point* of I if $y_j = x_j$ for $1 \leq j \leq k$ and $y_j = 0$ for all $j > k$. Also, if x' is the same as x except that its $(k+1)^{rst}$ component is the opposite of that of x, then we call the cells $N_{k+1}(x)$ and $N_{k+1}(x')$ the two *halves* of I.

Lemma 14.39 *Every cell $K \subseteq F^\infty$ satisfies one of the following possibilities: either*

$$K = F^\infty \text{ or } K = f_l^{-1}(I) \text{ for some } I \in \mathcal{F}_{l+1},\, l \geq 0, \tag{14.53}$$

or

$$\begin{aligned} &K \text{ is contained in a cell } L \text{ of the type (14.53)}, \\ &\text{and } K \text{ contains the left endpoint of exactly one} \\ &\text{of the two halves of } L. \end{aligned} \tag{14.54}$$

Indeed let K be any cell in F^∞, which we may as well assume is not F^∞ itself. We may also assume that K contains neither a nor b, because these are the left endpoints of the two halves of F^∞, and if K contains either of them then K satisfies (14.54).

With these assumptions we have that $K \subseteq f_0^{-1}(J)$ for some $J \in \mathcal{F}_1$. This is not hard to check, going back to the definitions of $\mathcal{F}_1$ and $f_0 = g_0 = \beta$. It comes down to the statement that K is a subset of either $A_k(a)$ or $A_k(b)$ for some k.

Choose $l \geq 0$ as large as possible so that $K \subseteq f_l^{-1}(I)$ for some $I \in \mathcal{F}_{l+1}$. If $K = f_l^{-1}(I)$, then K satisfies (14.53), and we are finished. If K contains the left endpoint of one of the two halves of $f_l^{-1}(I)$, then K satisfies (14.54), and we are again finished. Thus we assume that neither of these possibilities obtains. This assumption leads easily to the conclusion that $K \subseteq f_l^{-1}(I')$ for some $I' \in \mathcal{F}_{l+2}$. Indeed, $f_l^{-1}(I)$ is a cell on which f_l is rigid, and inside I $\mathcal{F}_1(I)$ looks exactly like $\mathcal{F}_1$ does inside of F^∞, and so the conclusion that $K \subseteq f_l^{-1}(I')$ for some $I' \in \mathcal{F}_{l+2}$ can be seen as a rephrasal of the observation that a cell properly contained in F^∞ which contains neither a nor b must lie inside $f_0^{-1}(J)$ for some $J \in \mathcal{F}_1$. The maximality of l forbids the inclusion of K in $f_l^{-1}(I')$ for such an I', and the lemma follows.

Given a pair of cells I and J, there is a natural bijection between I and J which is a combination of a backward shift (if needed, i.e., if they do not have the same size) and a rigid mapping. We call this the *normalized bijection* between the cells I and J.

Lemma 14.40 *If $L \subseteq F^\infty$ satisfies (14.53), then $f(L)$ is a cell. If $\phi : L \to F^\infty$ and $\psi : f(L) \to F^\infty$ are the normalized bijections from these cells onto F^∞, then the restriction of f to L is the same as $\psi^{-1} \circ f \circ \phi$.*

We may as well suppose that $L \neq F^\infty$, so that $L = f_l^{-1}(I)$ for some $I \in \mathcal{F}_{l+1}$, $l \geq 0$. Let $\psi : I \to F^\infty$ denote the normalized bijection from I onto F^∞. We know from Lemma 14.20 that f_l defines a rigid mapping from L onto I, and so ϕ is really the composition of ψ with the restriction of f_l to L.

On the other hand $f = h_l \circ f_l$, by definitions. We have that $h_l(I) = I$ (by Lemma 14.28), and therefore $f(L) = h_l(I) = I$ is a cell. Also, the restriction of h_l to $I = f_l(L)$ is really the same as $\psi^{-1} \circ f \circ \psi$. This can be derived from the definitions, especially the definition of the g_j's. Thus we get the desired representation for f on L, and the lemma follows.

Lemma 14.41 *Suppose that $K \subseteq F^\infty$ satisfies (14.54), and let L be the other cell promised in (14.54). Then there is another cell K' which satisfies (14.54) with $L' = F^\infty$ and normalized bijections ϕ from L onto F^∞ and ψ from $f(L)$ onto F^∞ such that ϕ maps K onto K' and the restriction of f to K is given by $\psi^{-1} \circ f \circ \phi$.*

This follows from the preceding lemma, applied to L. That is, we obtain ϕ and ψ first from the preceding lemma, and then we choose K' to be $\phi(K)$.

Lemma 14.42 *Let $k \geq 1$ be an integer and let c be an element of $\{a, b\}$. If η is the normalized bijection from $N_k(c)$ onto $N_1(c)$ and γ is the normalized bijection from $N_{k-1}(a)$ onto F^∞, then $f(N_k(c)) \subseteq N_{k-1}(a)$, and the restriction of f to $N_k(c)$ is the same as $\gamma^{-1} \circ f \circ \eta$ on $N_1(c)$.*

This can be derived from the definitions. Indeed, we have that

$$N_k(c) = \{c\} \cup \Big(\bigcup_{l=k+1}^{\infty} A_l(c) \Big). \tag{14.55}$$

Thus $f_0(N_k(c)) = \beta(N_k(c))$ is the union of some elements of $\mathcal{F}_1$ together with $f(c)$ $(= a)$, with $f_0(N_k(c)) \subseteq N_{k-1}(a)$. Lemma 14.28 (with $l = 0$) implies that $f(N_k(c)) = f_0(N_k(c))$, and so we get $f(N_k(c)) \subseteq N_{k-1}(a)$.

If γ is as in the lemma, then the elements of $\mathcal{F}_1$ in $f(N_k(c))$ are mapped to other elements of $\mathcal{F}_1$ under γ. Indeed, γ is just a backward shift, and we know exactly what the elements of $\mathcal{F}_1$ look like (14.25). Similarly the $A_l(c)$'s contained in $N_k(c)$ and $N_1(c)$ correspond to each other under η. The restrictions of f to each of the $A_l(c)$'s all look alike, they all correspond under normalized bijections, essentially by construction. It is not hard to show that the restrictions of f to $N_k(c)$ and $N_1(c)$ correspond as in the lemma, because the pieces match up. We leave the details to the reader.

The combination of the preceding lemmas says that the restriction of f to any cell in F^∞ is the same (modulo compositions on the left and on the right by normalized bijections) as the restriction of f to either F^∞ itself, $N_1(a)$, or $N_1(b)$.

Let us now summarize many of our conclusions.

Proposition 14.43 *The mapping $f : F^\infty \to F^\infty$ is 2-Lipschitz, preserves measure (in the sense of Lemma 14.37), and has BPI (as a mapping, as in Definition 14.11), but in each cell K there is a pair of points $x, y \in K$ such that $d(x, y) \geq \frac{1}{8} \operatorname{diam} K$ such that $f(x) = f(y)$.*

Indeed, the Lipschitz condition comes from Lemma 14.26, and the BPI condition follows from the preceding lemmas, as in the remark preceding the proposition. The behavior of β and f on $N_1(a)$ and $N_1(b)$ are not quite exactly the same, but certainly it is more than enough to get BPI. Indeed, for BPI one only needs to find large subsets in which things are practically the same. For that a simpler version of Lemma 14.42 would suffice, in which one says that the restrictions of f to the sets $A_k(a)$, $A_k(b)$, $k \geq 2$, look alike modulo compositions on the left and on the right by rigid mappings, and in fact they all look like f on F^∞. Thus we certainly have the BPI condition, but in fact we have much more self-similarity than that.

For the last part, about double points, we use Lemma 14.38. This says that any given cell K contains a cell of the form $f_l^{-1}(I)$, $I \in \mathcal{F}_{l+1}$, with at least $1/4$ the diameter of K. This gives double points in K because g_{l+1} has a pair of double points in I, and because f_j is rigid on $f_l^{-1}(I)$. The double points for g_{l+1} on I are at distance equal to $1/2$ of $\operatorname{diam} I$, because g_{l+1} is the same as β_I, and β_I is the same as β modulo composition on the right and left by normalized bijections.

We could have used Lemma 14.38 to get the BPI property of f, but the lemmas given afterwards help to bring out the self-similarity in a stronger way.

15

UNIFORMLY DISCONNECTED SPACES

15.1 Introduction

Definition 15.1 *Let $(M, d(x,y))$ be a metric space. We say that M is uniformly disconnected if there is a constant $C > 0$ so that for each $x \in M$ and $r > 0$ we can find a closed subset A of M such that $A \subseteq B(x,r)$, $A \supseteq B(x, C^{-1}r)$, and* $\operatorname{dist}(A, M \backslash A) \geq C^{-1} r$.

Notice that we can simply take $A = M$ when $r > \operatorname{diam} M$.

Roughly speaking, we are asking that there be an isolated "island" in M at any point and with practically arbitrary diameter whose distance to the rest of M is at least a fixed multiple of its diameter. Basic examples include our usual self-similar Cantor sets, as in Section 2.3. Later, when we discuss doubling measures, we shall discuss some amusing ways to deform the geometry of a Cantor set. For the moment let us merely point out that the set described in Section 11.6 is also uniformly disconnected.

If a space is uniformly disconnected then it is certainly totally disconnected, i.e., it contains no connected subsets with more than one element. Uniform disconnectedness can be seen as a scale-invariant version of total disconnectedness which comes with estimates. One can check that uniform disconnectedness is preserved under limits as in Chapter 8, at least if one has bounds on the uniform disconnectedness constants. One could also characterize uniform disconnectedness in terms of the total disconnectedness of the space and all its weak tangents, but we shall not bother. However, we shall touch on many of the relevant points in the next section.

Total disconnectedness is the same as "topological dimension 0", and one can also consider quantitative scale-invariant versions of "topological dimension $\leq n$" (or $= n$). This is discussed in more detail in [Se6], including relationships with taking limits of spaces.

Uniform disconnectedness may not seem like much fun, but we have seen already how even Cantor sets and their relatives can be pretty tricky. With uniformly disconnectedness we have a general way to make precise the idea of sets which are roughly like Cantor sets, and some way to see this class of sets as a whole. We shall see later that there is a kind of "uniformization theorem" for uniformly disconnected sets. When we talk about doubling measures we shall encounter some more questions about these sets which seem to be somewhat tricky and which are nicely connected to classical analysis.

Still uniform disconnectedness should not be as much fun as sets with lots of curves for instance. Uniformly disconnected sets should also be seen as just a place to start.

15.2 Spaces of dimension ≤ 1

Lemma 15.2 *If $(M, d(x,y))$ is a metric space which is Ahlfors regular of dimension $\alpha < 1$, then M is uniformly disconnected, with a constant which depends only on α and the Ahlfors regularity constants.*

To see this let $x \in M$ and $r > 0$ be given. Also let $\delta > 0$ be given, to be chosen soon, and small. According to Lemma 5.1, we can cover $B(x,r)$ by $\leq C\,\delta^{-\alpha}$ balls of radius $\delta\,r$, where C does not depend on x, r, or δ. If δ is small enough then we can find an interval $I \subseteq [r/2, r]$ of length $\delta\,r$ such that $d(x,y) \notin I$ for any $y \in M$. Otherwise we would be able to cover $[r/2, r]$ with $\leq C\,\delta^{-\alpha}$ intervals of length $\leq 4\,\delta\,r$, which is impossible if δ is small enough, because $\alpha < 1$. If s is the left endpoint of I, then $r/2 \leq s \leq r$ and $\operatorname{dist}(M \backslash B(x,s)) \geq \delta\,r$. This proves the lemma.

Proposition 15.3 *Let $(M, d(x,y))$ be an Ahlfors regular metric space of dimension 1 which is BPI. Then M is either uniformly disconnected or uniformly rectifiable.*

The idea here is that if M is not uniformly disconnected, then it should have a weak tangent which contains a nontrivial rectifiable arc. It is important here to have dimension 1 to get a *rectifiable* arc, since otherwise there could be a nontrivial connected subset of the weak tangent which is something like a snowflake.

Although this proposition is not deep, it does provide an amusing "classification" of BPI sets of dimension 1. One can imagine similar questions for dimensions strictly between 1 and 2, in which families of curves are allowed but nothing like "surfaces". It is not clear how to sort out the zoo of possibilities in a nice way.

In order to prove Proposition 15.3 we shall establish first the following auxiliary criterion for uniform disconnectedness.

Lemma 15.4 *Let $(M, d(x,y))$ be a metric space which is Ahlfors regular of dimension 1. Suppose that there is a constant $\eta > 0$ such that for each $x \in M$ and $r > 0$ there is a nonempty set $A \subseteq B(x,r)$ with the property that $\operatorname{dist}(A, M \backslash A) \geq \eta\,r$. Then M is uniformly disconnected.*

In other words, we assume the existence of isolated islands at all locations and scales in M, but we do not require that the island actually contain the given point x.

It is important here that M have dimension 1, because the lemma is false for spaces of dimension > 1. It is easy to build counterexamples in the plane, regular sets which contain a line segment but which are uniformly disconnected away from the line segment, with the uniformly disconnected parts accumulating on the segment in a reasonable manner. (See Figure 15.1.)

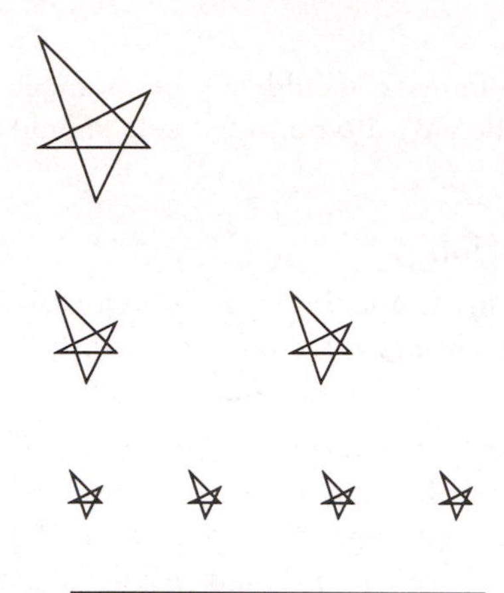

FIG. 15.1. A possible counterexample

To prove the lemma we argue by contradiction. Suppose that M and η are given as above but M is not uniformly disconnected. Let $\theta > 0$ be truly tiny, to be chosen soon. Let $x \in M$ and $r > 0$ be given, and assume that (x, r) is bad for the definition of uniform disconnectedness for this choice of θ. That is, we assume that there does not exist a set $E \subseteq B(x,r)$ such that $E \supseteq B(x, \theta\, r)$ and $\mathrm{dist}(E, M \backslash E) \geq \theta\, r$.

Let us pause for a definition (which we shall use later).

Definition 15.5 (Chains) *A finite sequence $\{z_i\}$ in M is called a γ-chain if $d(z_i, z_{i+1}) < \gamma$ for all i (except the last one).*

Think of γ as being small, so that a γ-chain is like an approximation to a curve.

Let us check that there is a $\theta\, r$-chain $x_1, \ldots, x_n$ in M such that $x_1 = x$ and $d(x_n, x) \geq r$. Indeed, if this were not the case, then we could take E to be the set of points which can be connected to x by a $\theta\, r$-chain, and this set E would be a set of the type that we are assuming does not exist (i.e., $B(x, \theta\, r) \subseteq E \subseteq B(x,r)$ and $\mathrm{dist}(E, M\backslash E) \geq \theta\, r$).

We may as well assume that our chain $x_1, \ldots, x_n$ also satisfies $d(x, x_i) < r$ when $i < n$, since otherwise we can throw away some terms of the sequence. Let Γ denote the subset of M consisting of the points $x_1, \ldots, x_n$. The idea is that Γ is very close to being a curve. The hypotheses of the lemma ensure that there are isolated islands at all locations and scales, and in particular which lie near Γ at various scales. These islands cannot get too close to Γ, because Γ is practically connected. By getting islands at many different scales this way we shall have too much mass in M, i.e., we shall violate Ahlfors regularity of dimension 1. To make this precise we show the following.

Sublemma 15.6 *Set $G(s) = \{z \in M : \mathrm{dist}(z, \Gamma) \in (\eta\, s, s)\}$ for $s > 0$, where η is as in Lemma 15.4. If θ is small enough and $\sqrt{\theta}\, r \leq s \leq r/10$, then*

$$H^1(G(s)) \geq C^{-1}\, r \tag{15.1}$$

where the constant C does not depend on x, r, s, or Γ. The smallness requirement on θ does not depend on x, r, s, or Γ either.

This is not too hard to prove. Given $y \in \Gamma$ and $0 < s \leq r/10$ we use the hypotheses of Lemma 15.4 to get a nonempty set $A(y,s) \subseteq B(y,s)$ which satisfies

$$\operatorname{dist}(A(y,s), M \backslash A(y,s)) \geq \eta\, s. \tag{15.2}$$

Thus $A(y,s)$ contains a ball of radius $\eta\, s$ and therefore

$$H^1(A(y,s)) \geq C^{-1}\, \eta\, s. \tag{15.3}$$

If $\sqrt{\theta}\, r \leq s$ and $\sqrt{\theta}$ is much smaller than η, then we may conclude that $A(y,s) \cap \Gamma = \varnothing$. Indeed, Γ is practically connected at the scale of $\eta\, r$, and if $A(y,s)$ intersected Γ then we would violate the isolation condition (15.2). Keep in mind our assumption that $s \leq r/10$, so that Γ is not stupidly small compared to the minimal isolation $\eta\, s$. Since $A(y,s)$ is disjoint from Γ but $y \in \Gamma$ we conclude that

$$A(y,s) \subseteq G(s) \tag{15.4}$$

when θ is small enough and $\sqrt{\theta}\, r \leq s \leq r/10$.

We would like the $A(y,s)$'s to be disjoint, and we can arrange that by choosing the y's carefully. That is, we choose points $y_j \in \Gamma$ such that $d(y_j, y_k) > 2\, s$ when $j \neq k$. We can choose a set of approximately r/s points y_j in Γ with this property, because of the assumptions on Γ. (For instance, we can choose them so that $d(y_j, x)$ is almost $3\, j\, s$, to within an error of about $\theta\, r$.) With this separation property we have that the sets $A(y_j, s)$ are pairwise disjoint. From here the desired lower bound on $H^1(G(s))$ follows easily, and the sublemma is proved.

Let us now derive Lemma 15.4 from the sublemma. For $0 < s \leq r/10$ we have that $G(s) \subseteq B(x, 2\, r)$, and we have that $G(s_1) \cap G(s_2) = \varnothing$ unless $\eta\, s_1 < s_2 < \eta^{-1}\, s_1$. We can choose approximately $\eta^2/\sqrt{\theta}$ values of s in the range $\sqrt{\theta}\, r \leq s \leq r/10$ which are pairwise separated in this manner, and we conclude from the sublemma that $H^1(B(x, 2\, r))$ is bounded from below by a multiple of $\eta^2/\sqrt{\theta}$. This contradicts Ahlfors regularity when θ is sufficiently small, and Lemma 15.4 follows.

The square root of θ was used only for notational convenience. One should really take θ to be a small multiple of η if one cares about the constants.

Let us now derive Proposition 15.3 from Lemma 15.4. Let M be a metric space which is Ahlfors regular of dimension 1 and which is not uniformly disconnected. We want to show that a weak tangent of M contains a nontrivial rectifiable curve.

Since M is not uniformly disconnected, we can find a sequence of points $\{x_j\}$ in M, a sequence of radii $\{r_j\}$, and a sequence of small positive numbers $\{\eta_j\}$ such that $\eta_j \to 0$ as $j \to \infty$ but (x_j, r_j) is bad for the definition of uniform disconnectedness with η taken to be η_j.

In fact we may assume that (x_j, r_j) is bad for the criterion for uniform disconnectedness provided by Lemma 15.4 and with η taken to be η_j. As a practical matter it is convenient to convert this to a statement about chains as before. That is, for each j we can find an $\eta_j\, r_j$-chain which connects x_j to $M \backslash B(x_j, r_j)$. This implies in particulat that $r_j \leq \operatorname{diam} M$.

Let M_{x_j, r_j} be as in Section 9.1. By passing to a subsequence we may assume that $\{M_{x_j, r_j}\}$ converges to a pointed metric space $(M', d'(\cdot,\cdot), x')$ which is a weak tangent of M. We know from Proposition 9.7 that M' is also Ahlfors regular of dimension 1.

From the convergence of the M_{x_j, r_j}'s it is easy to see that for j sufficiently large there is a $2\,\eta_j$-chain which connects x' to $M' \backslash B'(x', 1/2)$. We use $B'(\cdot,\cdot)$ to denote balls in M'. We may as well assume that this holds for all j, by throwing away the bad j's.

We want to use these chains to make a nontrivial rectifiable curve, but first we need to make some adjustments. We may assume that these chains are *minimal*, that one cannot remove any points from them without disturbing their important properties (of being $2\,\eta_j$-chains which connect x' to $M' \backslash B'(x', 1/2)$). Indeed, if any of them are not minimal, then we can simply throw away points to get minimal chains. Remember that these chains are finite sequences by definition.

Given a minimal η-chain $y_1, \ldots, y_n$ which connects x' to $M' \backslash B'(x', 1/2)$ in M', with $\eta > 0$ small, we have that $d'(y_l, y_{l+2}) \geq \eta$ for each l (except the last two) because of minimality. Thus the balls $B'(y_l, \eta/2)$ are pairwise disjoint if we restrict ourselves to even integers l. All of these balls are contained in $B'(x', 1)$, and we conclude that $n \leq C/\eta$ for a constant C that depends only on the regularity constant for M'.

Let us think of our minimal η-chain y_l as a mapping from the set $A = \{l/n : 1 \leq l \leq n\}$ into M. Our bound on n ensures that this mapping is C-Lipschitz where C depends only on the Ahlfors regularity constant for M'. Note that n is bounded from below by a constant multiple of η, and so $A \subseteq [0,1]$ is very thick in $[0,1]$ when η is small.

In our situation we have a sequence of minimal $2\,\eta_j$-chains and so we get a sequence of C-Lipschitz mappings $f_j : A_j \to M'$ which take the value x' at the first element of A_j and which take a value in $M' \backslash B'(x', 1/2)$ at the end of A_j. Moreover $A_j \subseteq [0,1]$ for each j, and

$$\lim_{j\to\infty} \sup_{t\in[0,1]} \operatorname{dist}(t, A_j) = 0. \tag{15.5}$$

This follows from the fact that the η_j's tend to 0. Using the Arzela–Ascoli argument we can pass to a subsequence and get a limiting mapping $f : [0,1] \to M'$ which is C-Lipschitz and satisfies $f(0) = x'$, $f(1) \in M' \backslash B'(x', 1/2)$.

Thus the weak tangent M' to M contains a nontrivial curve, as promised.

At this point we could say that the real line looks down on M' and conclude as in Proposition 11.10, but in this case there is a simpler argument which we provide instead. Define $g : M' \to \mathbf{R}$ by $g(z) = d'(z, x')$. This is a 1-Lipschitz function, and so $g \circ f : [0,1] \to \mathbf{R}$ is C-Lipschitz. We also have that $g(f(0)) = 0$ and $g(f(1)) \geq 1/2$. This implies that $g \circ f$ is bilipschitz on a set of positive measure, by standard results in real analysis. In fact, one can even get uniform bounds on the size of the set and on the bilipschitz constant, as in [D1, D4].

Now we use the assumption that M is BPI (for the first time). We know from Proposition 9.9 that M' is BPI and BPI equivalent to M. From Proposition 7.1 we get that M' is BPI equivalent to the real line, and hence M is too, by the transitivity of BPI equivalence (Proposition 7.5). We now apply Proposition 7.6 to conclude that M is uniformly rectifiable.

This completes the proof of Proposition 15.3.

15.3 Ultrametrics

Recall that a metric $d(\cdot,\cdot)$ on a set M is called an *ultrametric* if

$$d(x,z) \leq \max(d(x,y), d(y,z)) \tag{15.6}$$

for all triples of points $x, y, z \in M$. This is much stronger than the triangle inequality.

Proposition 15.7 *A metric space $(M, d(x,y))$ is uniformly disconnected if and only if there is a constant C and an ultrametric $d'(x,y)$ on M such that*

$$C^{-1}\, d(x,y) \leq d'(x,y) \leq C\, d(x,y). \tag{15.7}$$

This is probably at least implicit in the literature, and at any rate plenty of people must have known it already.

The proof is quite easy. If M is uniformly disconnected then we define $d'(x,y)$ to be the infimum of the $\gamma > 0$ such that there is a γ-chain (Definition 15.5) which starts at x and ends at y. This infimum is clearly $\leq d(x,y)$, and it is easy to see that the property of uniform disconnectedness implies that the infimum is $\geq C^{-1}\, d(x,y)$ for a suitable constant C. It is easy to check that this defines an ultrametric.

This construction can be viewed as a variation of the idea of the distance as the minimal length between curves.

For the converse we may as well assume that $d(x,y)$ is already an ultrametric, because the property of uniform disconnectedness is not disturbed by such a bounded perturbation of the metric. For an ultrametric we have that

$$\operatorname{dist}(B(x,r), M \backslash B(x,r)) \geq r \tag{15.8}$$

for each $x \in M$ and $r > 0$. This is easy to check, and the proposition follows.

For future reference let us record the following observation.

Lemma 15.8 *In an ultrametric space $(M, d(x,y))$ any pair of balls is either disjoint or one is contained inside the other.*

This is easy to verify from the definitions.

Remark 15.9 We know that in any Ahlfors regular space we can find a family of cubes as in Section 5.5, but on uniformly disconnected spaces we can get a nicer family of cubes. Namely, instead of the small boundary condition, we can simply have cubes for which the distance to their complement is bounded from below by a multiple of their diameter. Indeed, we may assume as in Proposition 15.7 that the space is equipped with an ultrametric. Then we can take for the cubes simply the collection of balls of radius 2^k for integers k. This has the required properties.

If we have two collections of cubes on a uniformly disconnected space which enjoy this separation property, then the two collections must be almost the same, in the sense that any cube in one collection can be realized as the union of a bounded number of cubes in the other collection of approximately the same size. This kind of "uniqueness" property does not work for other spaces, such as Euclidean spaces for instance.

15.4 Uniformization

In this section we want to give a kind of "uniformization" result, to the effect that uniformly disconnected sets can all be represented by a simple model, under a mild assumption anyway. For this result we shall use quasisymmetric mappings, whose definition we recall now.

Definition 15.10 (Quasisymmetric mappings) *Let M, N be metric spaces, with metrics $d_M(\cdot,\cdot)$, $d_N(\cdot,\cdot)$, respectively. A mapping $f : M \to N$ between them is said to be quasisymmetric if it is not constant and if there exists a homeomorphism $\eta : [0,\infty) \to [0,\infty)$ such that for any points $x, y, z \in M$ and $t > 0$ we have that*

$$d(x,y) \le t\, d(x,z) \quad \textit{implies that} \qquad\qquad (15.9)$$
$$\rho(f(x), f(y)) \le \eta(t)\, \rho(f(x), f(z)).$$

We may sometimes say that f is η-quasisymmetric to be more precise. A basic reference for quasisymmetric mappings is [TV].

Roughly speaking, quasisymmetric mappings distort *relative* distances by only a bounded amount. Bilipschitz mappings are automatically quasisymmetric, but the converse is not true. Bilipschitz mappings distort actual distances by only a bounded amount, whereas quasisymmetric mappings are permitted to distort actual distances in a strong way.

Notice that conformally bilipschitz mappings are quasisymmetric with a linear choice of η that does not depend on the scale factor. The mapping $f : \mathbf{R} \to \mathbf{R}$ defined by $f(x) = x\,|x|^{a-1}$ is quasisymmetric for any $a > 0$, but it is bilipschitz only for $a = 1$. One has similar examples on $\mathbf{R}^n$.

These examples point out an important feature of quasisymmetric mappings, which is that they distort distances by only a power, on reasonable spaces anyway (such as uniformly perfect spaces). Distortion by a power as opposed to exponential distortion or worse, for instance.

Another example: if $(M, d(x, y))$ is any metric space then the identity mapping is quasisymmetric as a mapping from $(M, d(x, y))$ to its snowflake transform $(M, d(x, y)^s)$. In particular quasisymmetric mappings can change Hausdorff dimensions.

Note that quasisymmetric mappings are always injective and continuous. The composition of two quasisymmetric mappings is again quasisymmetric, and the inverse of a surjective quasisymmetric mapping is also quasisymmetric. This is not difficult to verify. Two metric spaces are said to be *quasisymmetrically equivalent* if there is a quasisymmetric mapping from one onto the other.

Proposition 15.11 (Uniformization) *Suppose that $(M, d(x, y))$ is a compact metric space which is bounded, complete, doubling, uniformly disconnected, and uniformly perfect (Definition 5.3). Then M is quasisymmetrically equivalent to the symbolic Cantor set F^∞ (Section 2.3), where we take $F = \{0, 1\}$ and we use the metric $\rho(\cdot, \cdot)$ on F^∞ given by (2.4) with $a = 1/2$.*

The choice of the parameter a is irrelevant, in that all choices of $a \in (0, 1)$ correspond to snowflake transforms of the same metric and are quasisymmetrically equivalent to each other.

Proposition 15.11 applies in particular to the other symbolic Cantor sets in Section 2.3, based on finite sets with more than two elements. The fact that these sets are all quasisymmetrically equivalent is not too surprising, but remember that this does not work for bilipschitz equivalence, even when the Hausdorff dimensions match up, as in Section 11.9.

The assumption of boundedness here should be seen as merely a convenience. Completeness, the doubling condition, and uniform perfectness are essential, because they are necessary for the quasisymmetric equivalence with a Cantor set. Remember that boundedness implies compactness in the presence of doubling and completeness.

This proposition is not too surprising, and we should emphasize the likelihood that it has been observed before. It is simply a quantitative version of the classical fact that a compact totally disconnected perfect set is homeomorphic to the standard Cantor set. For our purposes it will be especially nice in connection with the story of doubling measures in the next chapter. Keep in mind though that this kind of "uniformization" does not work for bilipschitz equivalence, as in Proposition 11.18.

Let us now prove the proposition. Let M be as above, and let us assume that the metric $d(x, y)$ on M is actually an ultrametric, which we may do because of Proposition 15.7.

Lemma 15.12 *There is a constant C_0 so that for each ball B in M of radius r we can find a collection of balls B_i, $i = 1, 2, 3, \ldots, n$, $2 \leq n \leq C_0$, such that the B_i's are pairwise disjoint, their union is B, and they have the same radius which lies between $r/2$ and r/C_0. This constant depends only on the constants implicit in the assumptions on M.*

It is important here that there be at least 2 balls.

Let a be a small constant, to be chosen in a moment. The doubling property ensures that we can cover B by a bounded number of balls of radius $a\,r$. We may as well only use balls that touch B, and then Lemma 15.8 implies that we may assume that these balls B_i are pairwise disjoint and that their union is B. The remaining point is that there is at least 2 of them. This is true if we take a small enough, because of the requirement that M be uniformly perfect. This proves the lemma.

To prove the proposition we make a coding argument. Let W denote the set of all binary *words*, which is to say finite strings of 0's and 1's. By a *subword* we mean an initial substring of a given word, without gaps. Thus 0, 01 are considered as subwords of the word 010, but 1, 00 are not.

It is sometimes convenient to think of the empty set as being a word, a subword of everything, but we shall not consider it as an element of W.

One can think of a word in W as defining a proper subset of our Cantor set F^∞, namely the set of infinite binary strings which begin with the given word. These are the same as the balls in F^∞, except for F^∞ itself, which can be viewed as the ball which corresponds to the empty string. Subwords in W correspond to supersets in F^∞.

Given a set X with at least two elements, a *coding* of X means an injective mapping $\Gamma : X \to W$ with the following three properties. The first is that no word in the image of Γ have length greater than the number of elements in X. The second is that no word in the image of Γ should be a subword of any other element in the image of Γ. The third property is that every word of length equal to the number of elements of X should have one of the words in the image of Γ as a subword.

To understand what this means it is helpful to think in terms of subsets of F^∞. The first property means that we are associating to elements of X balls which are not too small. The second means that the balls associated to words in the image of Γ are pairwise disjoint, and the third says that their union is all of F^∞.

It is easy to see that we can define such a coding for any finite set X with at least two elements. We start by assigning the word 1 to one element, and then requiring that all remaining elements be associated to words that begin with 0. If there is only one remaining element, then we associate it to 0 and stop. If there is more than one, we associate the word 01 to another element, and then we require that all remaining elements be associated to words that begin with 00. Repeating this as necessary we get the desired coding.

It is more pleasant to think of this geometrically in terms of balls in F^∞. It is easy to see that we could have been more efficient in the preceding construction, making more subdivisions, and we could describe more efficient algorithms directly in terms of binary words. The cruder argument is adequate for our purposes and easier to read, and it is easy to improve it if one so desires.

Our third property of codings has the following consequence. If w is a word, then either w is a subword of one of the words used in the coding, or w contains one of those words as a subword. Indeed, if we extend w to a sufficiently long word w_1, then w_1 must contain one of the words u used in the coding as a subword, because of the third property above. This implies that either w is a subword of u or vice versa.

Next we define a certain collection of balls in M. Afterwards we define a mapping from these balls to words, and this will be used to define our uniformizing map.

We start with M itself, which we can view as the ball with arbitrary center and radius $\frac{3}{2}\operatorname{diam} M$. We apply Lemma 15.12 to get a bounded collection of subballs with the properties listed above. We then apply the lemma to each of these balls, and then to each of the new balls, and so forth forever. In this way we get a collection Δ of balls in M with the following properties. M is an element of Δ. Each element B of Δ has at least 2 but only boundedly many *children*. The children of B are disjoint elements of Δ which are properly contained in B and enjoy the property that all other elements of Δ properly contained in B are contained in one of the children of B. If $B \in \Delta$ and B' is a child of B, then

$$C_0^{-1} \operatorname{radius} B \leq \operatorname{radius} B' \leq \frac{1}{2} \operatorname{radius} B, \tag{15.10}$$

where C_0 is as above. If $B_0, B_1 \in \Delta$, then there is a unique $B \in \Delta$ such that B contains both B_0 and B_1 but no child of B contains both of them. B is called the minimal common ancestor of B_0 and B_1.

Now we want to define a mapping $\Gamma : \Delta \backslash \{M\} \to W$. To do this we proceed as follows. Given $B \in \Delta$ let X_B denote the collection of children of B, and let $\Gamma_B : X_B \to W$ be a coding for X_B. This exists, as above. We define $\Gamma : \Delta \backslash \{M\} \to W$ recursively as follows. If B is a child of M, then we set $\Gamma(B) = \Gamma_M(B)$. Once $\Gamma(B)$ is defined we define $\Gamma(B')$ for the children B' of B by taking the word $\Gamma(B)$ associated to B and adding the word $\Gamma_B(B')$ to it at the end. By repeating this process indefinitely we define Γ as a single-valued function on all of $\Delta \backslash \{M\}$.

Lemma 15.13 *If B_0, B_1 are elements of Δ with $B_0 \subseteq B_1$, then $\Gamma(B_1)$ is a subword of $\Gamma(B_0)$. Conversely, if $\Gamma(B_1)$ is a subword of $\Gamma(B_0)$, then $B_0 \subseteq B_1$.*

Indeed, let balls $B_0, B_1 \in \Delta$ be given. We may as well assume that $B_0 \neq B_1$. Let B denote the minimal common ancestor of B_0 and B_1. If $B_0 \subseteq B_1$, then $B = B_1$. In this case either $B_0 = B_1$ or $\Gamma(B_0)$ is defined by attaching nontrivial words to the end of $\Gamma(B_1)$, and $\Gamma(B_1)$ is a proper subword of $\Gamma(B_0)$. Similarly, $\Gamma(B_0)$ is a subword of $\Gamma(B_1)$ if B_1 is contained in B_0, with equality of the words only in the case of equality of the balls. In particular $B_0 = B_1$ if we also have that $\Gamma(B_1)$ is a subword of $\Gamma(B_0)$. Suppose finally that B_0 and B_1 are both properly contained in B, so that they are contained in distinct children B_0', B_1' of B. By construction neither of $\Gamma_B(B_0')$, $\Gamma_B(B_1')$ is a subword of the other, and therefore they disagree at some stage. The same must be true of $\Gamma(B_0')$ and $\Gamma(B_1')$, and for $\Gamma(B_0)$ and $\Gamma(B_1)$ as well. This means that neither of $\Gamma(B_0)$ and $\Gamma(B_1)$ is a subword of the other in this case, and the lemma follows.

Our mapping Γ is not surjective, but it almost is.

Lemma 15.14 *If w is a word in W, then there are balls $B, B' \in \Delta$ such that B' is a child of B, $\Gamma(B)$ is a subword of w, and w is a subword of $\Gamma(B')$. For this we allow $B = M$, with $\Gamma(M)$ interpreted as the empty word. The difference between the lengths of $\Gamma(B)$ and $\Gamma(B')$ is uniformly bounded.*

Indeed, let $w \in W$ be given. The collection of balls B such that $\Gamma(B)$ is a subword of w is finite and linearly ordered by inclusion, by Lemma 15.13. Take B to be the minimal such ball. We need to show that w is a subword of $\Gamma(B')$ for some child B' of B. We may as well assume that $w \neq \Gamma(B)$. Let w_0 be the part of w which comes after the subword $\Gamma(B)$. From our story about codings we know that there is a child B' of B such that either w_0 is a subword of $\Gamma_B(B')$ or $\Gamma_B(B')$ is a subword of w_0. This means that either w is a subword of $\Gamma(B')$ or $\Gamma(B')$ is a subword of w. The latter possibility is ruled out by the minimality of B, and so we get the former, as desired.

Let us check the last statement in the lemma. We know that each set of children X_B has only a bounded number of elements, and the lengths of the words in the image of each Γ_B is bounded by the number of elements in X_B. This gives the bound on the difference in the lengths, and the lemma follows.

We can now define a mapping $h : M \to F^\infty$. Given $x \in M$, consider $\Delta_x = \{B \in \Delta : x \in B\}$. Because of the nesting properties of balls this is a linearly ordered set of balls. The words $\Gamma(B)$, $B \in \Delta_x$ are linearly ordered by the the subword relation, because of Lemma 15.13. Thus the collection Δ_x actually defines a unique infinite binary string, an element of F^∞. This uses also the fact that the length of $\Gamma(B)$ goes to infinity as the radius of B goes to zero. We take $h(x)$ to be this string.

Conversely, suppose that we are given an infinite binary string $s \in F^\infty$. Consider the collection $\mathcal{B}(s)$ of balls $B \in \Delta$ such that $\Gamma(B)$ is an initial finite substring of s. Lemma 15.14 implies that this collection contains balls of arbitrarily small size. These balls are linearly ordered by inclusion, by Lemma 15.13, and so they shrink down to a unique point x. Balls are closed sets, as in (15.8), and so every element of $\mathcal{B}(s)$ contains x. Thus $\mathcal{B}(s) \subseteq \Delta_x$. This means that s is actually the string that we have associated to x. Every element of F^∞ therefore lies in the image of h and determines its preimage uniquely.

Thus we get a one-to-one correspondence between M and F^∞. It is not hard to check that this mapping is quasisymmetric, using (15.10) and Lemma 15.14, but it is a bit tedious. We omit the details.

This completes the proof of Proposition 15.11.

15.5 Making regular mappings

Proposition 15.15 *Let $(M, d(x,y))$ be a bounded Ahlfors regular metric space of dimension γ. Then there is a uniformly disconnected metric space $(M_1, d_1(x,y))$ and a mapping $f_1 : M_1 \to M$ such that M_1 is also Ahlfors regular of dimension γ, f_1 is regular and surjective, and there is a σ-compact subset E_1 of M_1 of measure 0 such that f_1 is injective on the complement of E_1.*

The construction will show that E_1 is smaller than measure 0, that it has Hausdorff dimension $< \gamma$, for instance. The mapping is arguably better than we have stated too.

In principle Proposition 15.15 provides a way of saying that every space can be obtained from one which is uniformly disconnected by simply pushing some of the islands together. Thus uniformization for uniformly disconnected spaces provides information about all metric spaces.

In Proposition 16.9 we shall strengthen our uniformization result for uniformly disconnected spaces in the Ahlfors regular case.

Problem 15.16 *If, in Proposition 15.15, M is BPI, can we make M_1 BPI also? Can we make f_1 a BPI mapping?*

Our construction does not preserve symmetry in general.

Consider the case of the Heisenberg group, for instance. Think of starting with a compact regular subset of the Heisenberg group of maximal dimension, like the closure of a cube. Is there a corresponding Heisenberg Cantor set with special features? One could also ask about Heisenberg versions of other fractals.

Note that for cubes in a Euclidean space we can find BPI versions of M_1 and f_1 as in the proposition, using binary Cantor sets and binary decimal representations of real numbers, as in Section 11.3. We could also use different Cantor sets.

Let us prove the proposition. The idea is very simple. We start with a collection Δ of "cubes" on M as in Section 5.5. We may as well assume that M itself is an element of Δ. We shall build M_1 in such a way that it has practically the same collection of cubes, but we change the geometry by separating the cubes. We first make the combinatorial construction, and then choose a metric afterwards.

Let us call a subset X of Δ a *ray* if it satisfies the following properties: X contains the top cube M; X is linearly ordered by inclusion, so that $Q, T \in X$ implies either $Q \subseteq T$ or $T \subseteq Q$; if $Q \in X$, $T \in \Delta$, and $Q \subseteq T$, then $T \in X$; and X contains cubes of arbitrarily small diameter. More concretely every element of X should have exactly one child in X, and the parent of every element in X besides M should also lie in X.

Let $\mathcal{R}$ denote the collection of all rays in Δ. As a set M_1 will be taken to be $\mathcal{R}$, but we keep the separate notation because $\mathcal{R}$ has an interesting life of its own.

Lemma 15.17 *If two rays X and X' are different, then there are disjoint cubes $Q \in X$ and $Q' \in X'$.*

Indeed, if $X \neq X'$, then either there is a cube $Q \in X$ not in X', or vice versa, which is the same after interchanging the roles of X and X'. Any cube Q' in X' of smaller size than Q must be disjoint from Q, because otherwise $Q' \subseteq Q$ and then $Q \in X'$ too. This proves the lemma.

There is a natural mapping from $\pi : \mathcal{R} \to M$ defined as follows. If X is a ray in Δ, then

$$\bigcap_{Q \in X} \overline{Q} \tag{15.11}$$

is a nonempty subset of M with diameter 0, i.e., a set with exactly one element. That is the point $\pi(X)$.

Every element of M can arise this way. Indeed, given $x \in M$, let X denote the collection of cubes in Δ which contain x. This is a ray because of the nesting properties of cubes.

A point in M can come from more than one ray in general, because the point might lie on the boundary of a pair of cubes. That is the only way that multiplicities can occur. Let us state this more formally.

Lemma 15.18 *There is a σ-compact set $E \subseteq M$ of measure 0 such that if $x \in M \backslash E$, then there is only one ray X such that $\pi(X) = x$.*

For this we take E to be the union of all the intersections of the closures of disjoint cubes, i.e.,

$$E = \bigcup \{\overline{Q}_1 \cap \overline{Q}_2 \, : \, Q_1, Q_2 \in \Delta, \, Q_1 \cap Q_2 = \varnothing\}. \tag{15.12}$$

This set has measure 0 because of the properties of cubes described in Section 5.5. That is to say, we use (5.9), and one can even show that E has Hausdorff dimension $< \gamma$.

Suppose now that $x \in M$ arises as both $\pi(X)$ and $\pi(X')$ for distinct rays X and X'. Then there exist cubes $Q \in X$ and $Q' \in X'$ which are disjoint, as noted above, and x must lie in the intersection of their closures. This implies that $x \in E$, and the lemma follows.

We shall be interested in some "cubical sets" in $\mathcal{R}$. Given a cube Q, define its lifting $\Lambda(Q)$ to $\mathcal{R}$ by

$$\Lambda(Q) = \{X \in \mathcal{R} : Q \in X\}. \tag{15.13}$$

Notice that $Q \subseteq \pi(\Lambda(Q)) \subseteq \overline{Q}$ for all cubes Q. Also, if $\{Q_i\}$ is any finite covering of M by cubes in Δ, then $\{\Lambda(Q_i)\}$ covers $\mathcal{R}$. This is easy to check.

These cubical sets provide a kind of combinatorial structure on $\mathcal{R}$, but now we want to define a more precise geometry.

Set

$$\rho(X, Y) = \inf\{\operatorname{diam} Q : Q \in X \cap Y\}. \tag{15.14}$$

Note that we can always take $Q = M$, and in particular

$$\rho\text{-}\operatorname{diam} \mathcal{R} \leq \operatorname{diam} M. \tag{15.15}$$

It is easy to see that $\rho(X, Y)$ is indeed a metric. It is obviously symmetric and nonnegative, and it vanishes exactly when $X = Y$. In fact $\rho(X, Y)$ is an ultrametric, as one can check. The balls for this metric are all sets of the form $\Lambda(Q)$, and the converse is almost true. (The problem is that a cube could have a proper subcube of the same diameter, which causes trouble.)

We have some easy estimates for $\rho(X, Y)$. The first is that

$$d(\pi(X), \pi(Y)) \leq \rho(X, Y) \quad \text{for all } X, Y \in \mathcal{R}, \tag{15.16}$$

which follows directly from the definitions. Next

$$\rho\text{-diam}\,\Lambda(Q) \leq \operatorname{diam} Q \tag{15.17}$$

for all $Q \in \Delta$, again by definitions.

Lemma 15.19 $\pi : \mathcal{R} \to M$ *is regular.*

We have that π is 1-Lipschitz, and so it suffices to show that the inverse image of a ball is covered by a bounded number of balls of the same radius. In fact it suffices to show that if $Q \in \Delta$ then $\pi^{-1}(Q)$ is contained in the union of $\Lambda(Q_i)$ for a bounded number of Q_i's of approximately the same size as Q. Let us take the Q_i's to be the cubes of about the same size as Q whose closures intersect the closure of Q. There are only boundedly many of these, and $\pi^{-1}(Q)$ is contained in their union because $\pi(T)$ is contained in the closure of T whenever $T \in \Delta$. This proves the lemma.

Lemma 15.20 $\mathcal{R}$ *is compact.*

Let $\{X_j\}$ be any sequence of rays, and let us show that there is a convergent subsequence. For each $\epsilon > 0$ there is a cube $Q \in \Delta$ with $\operatorname{diam} Q < \epsilon$ so that Q lies in X_j for infinitely many j's. Using a Cantor diagonalization argument we can find a subsequence of $\{X_j\}$ so that for each $\epsilon > 0$ there is a cube $Q \in \Delta$ with $\operatorname{diam} Q < \epsilon$ such that Q lies in all but finitely many of the X_j's. This is the subsequence that we want, let us call it $\{Y_k\}$.

Let Y consist of the cubes which are contained in Y_k for all but finitely many k's. Thus $M \in Y$, and it is easy to see that the elements of Y are linearly ordered by inclusion, because of the corresponding property of the Y_k's. Given any cube in Y, all the cubes which contain it must also be contained in Y, because of the corresponding property of the Y_k's. Also Y contains cubes of arbitrarily small diameter, by our choice of $\{Y_k\}$. This implies that Y lies in $\mathcal{R}$. It is not hard to see that $\{Y_k\}$ converges to Y with respect to the metric $\rho(\cdot, \cdot)$.

Lemma 15.21 $\mathcal{R}$ *is Ahlfors regular of dimension* γ.

It suffices to show the usual bounds for the Hausdorff measure of balls. The upper bound follows from the regularity of the mapping π and the Ahlfors regularity of M, as in Lemma 12.3. For the lower bound we observe first that

$$H^\gamma(\Lambda(Q)) \geq H^\gamma(Q). \tag{15.18}$$

We have this because $\pi(\Lambda(Q)) \supseteq Q$ and π is 1-Lipschitz. Suppose now that we have a ball in $\mathcal{R}$ with center X and radius $r \leq \operatorname{diam} \mathcal{R}$. This means that $r \leq \operatorname{diam} M$, and we can find a cube $Q \in X$ such that $C^{-1} r \leq \operatorname{diam} Q < r$, where C does not depend on X or r. We have that $X \in \Lambda(Q)$ and that $\Lambda(Q)$ is contained in the given ball, because of (15.17). Thus the H^γ measure of the ball is bounded from below by $H^\gamma(\Lambda(Q))$ and hence by $H^\gamma(Q)$, which implies the desired lower bound. This proves the lemma.

One could approach the lemma differently, trying to make an Ahlfors regular measure on $\mathcal{R}$ by pulling back one on M, but the preceding is a little more convenient technically.

Lemma 15.22 *There is a σ-compact set $E_1 \subseteq \mathcal{R}$ of measure 0 such that π is injective on $\mathcal{R}\backslash E_1$.*

For this it suffices to take $E_1 = \pi^{-1}(E)$, where E is as in Lemma 15.18. This set has measure 0 because of Lemma 12.3, since π is regular. The σ-compactness uses the compactness of $\mathcal{R}$.

We are now finished, except for renaming $(\mathcal{R}, \rho(\cdot,\cdot))$ as $(M_1, d_1(\cdot,\cdot))$, π as f_1, and observing that this space is uniformly disconnected because $\rho(\cdot,\cdot)$ is an ultrametric, as observed above. Remember also Proposition 15.7.

This completes the proof of Proposition 15.15.

Remark 15.23 (About cubes) The sets $\Lambda(Q)$, $Q \in \Delta$, are pairwise disjoint and provide a collection of cubes for $\mathcal{R}$ in the sense of Section 5.5. In fact the balls for $\mathcal{R}$ with respect to the ultrametric $\rho(\cdot,\cdot)$ are $\Lambda(Q)$'s, and the converse is nearly true, as we mentioned above. At any rate the $\Lambda(Q)$'s enjoy practically the same properties as discussed in Section 15.3.

Remark 15.24 (Other geometries) We can always find a uniformly disconnected "covering", as in the proposition, but what about other kinds of geometries? For instance, if a space has topological dimension greater than 1, one might be interested in coverings of topological dimension 1. One might want this covering to be modelled on a snowflake curve, or maybe a fractal tree, or the product of a uniformly disconnected set with an interval. The possibilities may depend on the geometry of the space. One can try to analyze this and make constructions as above. We have merely investigated the simplest case of uniform disconnectedness, but there is a zoo of other scenarios to consider. Here too there is the issue of putting self-similarity back into the picture.

Remark 15.25 (Making sets from nested cubes) In the context of the construction described in this section one should remember the special "universe" of sets made from nesting cubes as in Section 2.5 and Chapter 13. These sets come with natural families of cubes built into the construction. In some cases these cubes will not quite satisfy the requirements of Section 5.5, because of problems with the condition (5.9) concerning small boundaries, but the combinatorics remain the same. In this case the idea of moving cubes apart is much easier to understand, and can be implemented more simply in the same framework. (It may be helpful to work in a larger Euclidean space to make it easier to move the cubes apart.)

Remark 15.26 (Adding connections) In this section we made a construction for realizing a given space as the image under a regular mapping of a space which was "simpler" in the sense of being less connected. One can instead try to go in the opposite direction and "add connections" to make a regular mapping from the given space onto one which is more connected. Moving cubes together instead of pulling them apart. Given an Ahlfors regular space of dimension $\geq n$, under what conditions can one make a regular mapping onto an Ahlfors regular space of topological dimension n? There is also the problem of being able to do this in a way which respects some self-similarity, like BPI. One can begin concretely with sets constructed as in Section 2.5 and Chapter 13.

16

DOUBLING MEASURES AND GEOMETRY

16.1 The definition

Definition 16.1 *Let $(M, d(x, y))$ be a metric space. A positive Borel measure μ on M is said to be* doubling *if there is a constant $C > 0$ such that*

$$\mu(2\,B) \leq C\,\mu(B) \tag{16.1}$$

for all balls B in M (and if μ is not identically zero).

Note that $\mu(B)$ is always positive when μ is doubling and B is a ball.

We shall discuss examples soon, but notice that Hausdorff measure of the correct dimension on any Ahlfors regular metric space is doubling. We shall often be interested in other doubling measures on Ahlfors regular spaces, for which we think of the doubling measure as some kind of controlled deformation of Hausdorff measure.

Notice that if μ is a doubling measure on M and μ' is a doubling measure on M', then $\mu \times \mu'$ is a doubling measure on $M \times M'$.

In this section we record some basic facts.

Lemma 16.2 *Suppose that $(M, d(x, y))$ is a metric space and that μ is a doubling measure on M. Then*

$$\mu(2^k\,B) \leq C^k\,\mu(B) \tag{16.2}$$

for some constant C, all balls B, and all positive integers k. If M is uniformly perfect (Definition 5.3), then there is a (small) constant $0 < a < 1$ such that

$$\mu(a^k\,B) \leq (1-a)^k\,\mu(B) \tag{16.3}$$

for all balls B and all positive integers k.

The first part is an immediate consequence of the definitions. The second is not hard to check, and we leave it as an exercise. An argument is given in [Se8].

In practice we shall be happy to assume that our spaces are uniformly perfect, even though that is not necessary for many purposes. Counting measure on the integers is a perfectly good doubling measure, and while it does not satisfy the decay property above, it does enjoy this property at scales larger than 1.

Note that the decay property above implies that the measure has no atoms. In general a doubling measure can have atoms only at isolated points in the space.

Lemma 16.3 *If $(M, d(x, y))$ is a metric space which admits a doubling measure, then M is doubling as a metric space (Definition 5.5).*

This can be proved in the same manner as Lemma 5.1.

A consequence of this lemma is that closed and bounded subsets of M are compact if M is complete and admits a doubling measure. In this case any doubling (or locally finite) measure on M is Borel regular, because of Theorem 2.18 in [R].

Lemma 16.4 *Let M and N be metric spaces. If $f : M \to N$ is a quasisymmetric bijection (Definition 15.10), then doubling measures on M and N correspond to each other under f.*

This is not too hard to check, using (16.2) and the definitions. A special case of this lemma is that the class of doubling measures on a metric space is not changed by the snowflake transform.

16.2 Deformations of geometry

Suppose that $(M, d(x, y))$ is a metric space and that μ is a doubling measure on M. Let $\alpha > 0$ be given, and set

$$D(x, y) = \{\mu(\overline{B}(x, d(x, y))) + \mu(\overline{B}(y, d(x, y)))\}^{\alpha}. \tag{16.4}$$

We think of this as defining a deformation of the geometry of M. If M is n-dimensional Euclidean space with the Euclidean metric, μ is Lebesgue measure, and $\alpha = 1/n$, then $D(x, y)$ is just a constant multiple of the Euclidean metric. If M is an Ahlfors regular space of dimension n, μ is Hausdorff measure H^n, and $\alpha = 1/n$, then $D(x, y)$ is bounded above and below by constant multiples of the original metric. In general we can get a nontrivially different geometry.

Note that this definition of $D(x, y)$ does not rely too heavily on the original metric $d(x, y)$. The collection of balls determined by $d(x, y)$ matters a great deal, but their precise radii is less important. For instance, (16.4) is not changed under a snowflake transform $d(x, y) \mapsto d(x, y)^s$ on the original metric. More generally, (16.4) cooperates well with quasisymmetric deformations of $d(x, y)$.

Lemma 16.5 *For any doubling measure μ, $D(x, y)$ is a quasimetric.*

This is not too hard to check, and we omit the details. Note well that it need not be a metric, nor even comparable in size to a metric. Indeed, one could have $\mu = |x_1|\, dx$ on $\mathbf{R}^2$, where dx denote Lebesgue measure. In this case it is natural to define $D(x, y)$ with $\alpha = 1/2$, so that $D(x, y)$ is a relatively modest perturbation of the Euclidean metric, but then one has

$$D(x, y) = c\, |x - y|^{3/2} \tag{16.5}$$

when $x_1 = y_1 = 0$. Thus the restriction of $D(x, y)$ to the $x_1 = 0$ axis is very far from being a metric, Lipschitz functions on it must be constant, for instance.

Recall from (2.3) that any quasimetric has a positive power which is comparable in size to a metric (bounded above and below by positive multiples of a metric). In our case this means that we can always get something which is comparable in size to a metric by taking α sufficiently small. Typically we shall want to take α to be the reciprocal of the Hausdorff dimension, and it is not so natural to shrink α brutally.

In some cases we do not have this problem. On the real line, for instance, it is natural to take $\alpha = 1$, and in this case $D(x, y)$ is bounded from above and below by constant multiples of

$$D'(x, y) = \mu([x, y]). \tag{16.6}$$

It is easy to see that this is a true metric, and in fact that $(\mathbf{R}, D'(x, y))$ is isometrically equivalent to the real line with the standard metric. The isometry is obtained by integrating μ. There is nothing like this for $\mathbf{R}^n$ with $n > 1$, however. This is not too surprising, there is no simple way to make mappings from measures by "integrating". This is also related to the absence of arclength parameterizations in higher dimensions. See [Se8] for more discussion of these matters.

There is another case in which we do not have this problem.

Lemma 16.6 *Suppose that $(M, d(x, y))$ is an ultrametric space (as in Section 15.3) and that μ is a doubling measure on M. Let $D(x, y)$ be defined as in (16.4). Then $D(x, y)$ and $d(x, y)$ have the same collection of balls, as sets, and $D(x, y)$ is an ultrametric.*

To see this we start by observing that

$$\overline{B}(x, d(x, y)) = \overline{B}(y, d(x, y)) \tag{16.7}$$

since we have an ultrametric. Let us check that the D-ball

$$\{z \in M : D(x, z) \leq D(x, y)\} \tag{16.8}$$

is the same as the original ball $\overline{B}(x, d(x, y))$. That is, they are the same as sets, $D(\cdot, \cdot)$ and $d(\cdot, \cdot)$ may assign them completely different radii. This assertion is equivalent to

$$D(x, z) \leq D(x, y) \quad \text{if and only if} \quad d(x, z) \leq d(x, y). \tag{16.9}$$

Suppose first that $d(x, z) \leq d(x, y)$. Then

$$\overline{B}(x, d(x, z)) = \overline{B}(z, d(x, z)) \subseteq \overline{B}(x, d(x, y)), \tag{16.10}$$

where the first equality is like the one above and the inclusion follows from $d(x, z) \leq d(x, y)$. This implies that $D(x, z) \leq D(x, y)$. Conversely, if $d(x, z) > d(x, y)$, then it means that

$$\overline{B}(x, d(x, z)) = \overline{B}(z, d(x, z)) \supseteq \overline{B}(x, d(x, y)), \tag{16.11}$$

and the inclusion is proper because z does not lie in the latter ball. In fact the ultrametric property ensures that

$$B(z, d(x, y)) \subseteq \overline{B}(z, d(x, z)) \backslash \overline{B}(x, d(x, y)). \tag{16.12}$$

Therefore $\mu(\overline{B}(z, d(x, z))) > \mu(\overline{B}(x, d(x, y)))$, since $\mu(B(z, d(x, y))) > 0$. In this case one gets $D(x, z) > D(x, y)$, as desired. Thus we get (16.9), and D-balls are the same as d-balls as sets.

From here we conclude that the D-balls satisfy the same nesting properties as the d-balls (any pair of them is either disjoint or one is contained in the other), and the ultrametric property for D follows. This proves the lemma.

In general it is not true that the D-balls are the same as the d-balls. Under reasonable conditions they are approximately the same in shape even if they may be very different in size.

Lemma 16.7 *Suppose that $(M, d(x, y))$ is a metric space which is uniformly perfect (Definition 5.3) and that μ is a doubling measure on M. Let $D(x, y)$ be defined as in (16.4). Then the identity mapping on M, viewed as a mapping from $(M, d(x, y))$ to $(M, D(x, y))$, is quasisymmetric (Definition 15.10).*

Note that we defined quasisymmetric mappings originally only for metric spaces (rather than quasimetric spaces), but that does not really matter.

The lemma is not very difficult to prove, using Lemma 16.2, and we leave the details as an exercise. It is discussed in more detail in [Se8].

Lemma 16.7 is pleasant for the way in which it says that the deformation of a metric by a doubling measure is not too severe. Note that the assumption of uniform perfectness is not essential, and the same result holds on the $\mathbf{Z}$ equipped with the usual metric, for instance.

Notice also that the quasisymmetric equivalence provided by the lemma gives topological equivalence too.

Lemma 16.8 *Same assumptions as in Lemma 16.7. If $(M, d(x, y))$ is complete, then so is $(M, D(x, y))$, and the latter is Ahlfors regular of dimension $1/\alpha$, where α is as in the definition (16.4) of $D(x, y)$.*

Again for this we need to have an extended notion of Ahlfors regularity for spaces with quasimetrics instead of metrics. This is not a big deal, but there is a subtlety, which is that the balls with respect to quasimetrics need not be measurable *a priori*. This problem can be avoided in general by replacing the given quasimetric with a power of a metric, as in (2.3). In our situation there is a more pleasant resolution of this difficulty, which is that $D(x, y)$ is Borel measurable, and in fact semicontinuous. This can be checked using the simple fact that

$$\lim_{t \to r+} \mu(\overline{B}(x, t)) = \mu(\overline{B}(x, r)) \tag{16.13}$$

for all $x \in M$ and $r > 0$. This is true for any positive measure which is finite on bounded sets, and it implies the semicontinuity of $D(x, y)$.

Lemma 16.8 is also a fairly straightforward consequence of the definitions, and we omit the proof. One must be a little careful though, and an argument is provided in [Se8].

Thus we see that the deformations of metrics that come from doubling metrics behave rather well, at least under mild assumptions. Of course it remains to talk about examples and more concrete situations.

16.3 Uniformization revisited

Proposition 16.9 *Let $(M, d(x,y))$ be a compact Ahlfors regular metric space of dimension γ which is uniformly disconnected (Definition 15.1). Let F be the set $\{0,1\}$, and let F^∞ be as in Section 2.3. Then there is a doubling measure μ on F^∞ such that $(M, d(x,y))$ is bilipschitz equivalent to $(F^\infty, D(\cdot,\cdot))$, where $D(\cdot,\cdot)$ is defined as in (16.4) with $\alpha = 1/\gamma$.*

We should say here that we can equip F^∞ with a metric $d_a(x,y)$ with $0 < a < 1$, as in Section 2.3. These metrics are all related to each other by snowflake transforms, and so they all have the same class of doubling measures. Thus we may speak simply of "doubling measures on F^∞". Also, the choice of a does not affect the resulting $D(\cdot,\cdot)$, as one can easily check.

To prove the proposition we use Proposition 15.11. Let M be as above, and note that Ahlfors regularity implies the doubling and uniformly perfect conditions (by Lemmas 5.1 and 5.4). Proposition 15.11 provides a quasisymmetric equivalence $\phi : M \to F^\infty$, where for F^∞ one takes the metric $d_a(x,y)$ with $a = 1/2$ but this specific choice is irrelevant. Let μ be the measure on F^∞ that corresponds to Hausdorff measure H^γ on M under ϕ. μ is doubling because H^γ is doubling on M (by Ahlfors regularity) and because ϕ is quasisymmetric, as in Lemma 16.4. If we define $D(\cdot,\cdot)$ as in (16.4) with $\alpha = 1/\gamma$, then

$$C^{-1}\, d(x,y) \le D(\phi(x),\phi(y)) \le C\, d(x,y) \tag{16.14}$$

for all $x, y \in M$. This is not very hard to check, using quasisymmetry and Ahlfors regularity, and we leave the details as an exercise. This proves Proposition 16.9.

Thus in principle questions about the geometry of Ahlfors regular spaces which are uniformly disconnected are reduced to questions about doubling measures on F^∞.

16.4 Doubling measures on Cantor sets

Let F be a finite set with at least two elements, and let F^∞ be as in Section 2.3. We are interested in looking at doubling measures on F^∞ and the associated $D(\cdot,\cdot)$'s. For this we use a metric $d_a(x,y)$ on F^∞ as in Section 2.3, where the precise choice of $0 < a < 1$ does not matter for the present purposes.

Note that $d_a(x,y)$ is an ultrametric, and so the $D(\cdot,\cdot)$'s associated to doubling measures are too, as in Lemma 16.6.

The most basic examples of doubling measures on F^∞ – after the "uniform" measure – are defined as follows. Let λ be a probability measure on F which does not vanish on any of the elements of F, and take the infinite product of copies of λ to get a probability measure Λ on F^∞. It is not hard to check that Λ is doubling.

Let $D(x, y)$ be as defined in (16.4) using the doubling measure Λ, with $\alpha = 1$, say. Thus we get an ultrametric space $(F^\infty, D(x, y))$ which is Ahlfors regular of dimension 1, as in Lemmas 16.6 and 16.8.

For this choice of Λ, $(F^\infty, D(x, y))$ is a BPI space. This is easy to check. In fact all balls in $(F^\infty, D(x, y))$ are conformally isometrically equivalent to $(F^\infty, D(x, y))$, essentially by definitions. Lemma 16.6 is also relevant here.

If F has n elements, then we obtain in this way an $(n-1)$-parameter family of BPI spaces.

Problem 16.10 *When are these BPI spaces BPI equivalent?*

This question is interesting already for $n = 2$.

It would be quite nice to have nontrivial continuous families of BPI equivalence classes arising in this manner. So far our examples have had a kind of "discreteness" to them, in the range of choices that is. Practically the only way so far to make continuous families which we know to be nontrivial would be to vary snowflake parameters, as in Section 11.8, for example. (That is, one would vary *internal* snowflake parameters, taking care that the Hausdorff dimensions remain the same.)

There is a recurring theme in geometry, that it is easier to build continuous families of geometries in low topological dimensions than in high topological dimensions. The objects that we are looking at correspond to something like a sphere at infinity in the more typical discussions. The sphere at infinity of a discrete group maybe, or of the universal covering of some compact space. It would be interesting to see how this phenomena might manifest itself in the context of BPI spaces, where it is easier to make families of BPI spaces which are uniformly disconnected than families which have topological dimension 2, for instance. Here "families" should mean continuous families somehow, although we presently lack an adequate notion for that. See Chapter 17 for more details concerning our lack of success with continuous families. For the present purpose one would probably want a strong notion of continuous deformation, with uniform control over all scales.

We can make many more doubling measures than just infinite products of a single measure. Let $\mathcal{P}$ denote the set of probability measures on F. This is the same as collections of n nonnegative numbers whose sum is 1, where n is again the number of elements of F. Given $\epsilon > 0$ let $\mathcal{P}_\epsilon$ denote the set of probability measures on F which give mass at least ϵ to each element of F. Suppose that $\{\lambda_j\}$ is a sequence in $\mathcal{P}_\epsilon$ for some fixed ϵ, and let Λ be the product probability measure on F^∞. It is not very difficult to check that Λ is doubling.

If we define $D(x, y)$ on F^∞ using Λ, then it is not so clear when the space $(F^\infty, D(x, y))$ will be BPI. There are some obvious sufficient conditions, like the periodicity of $\{\lambda_j\}$, or mild errors from periodicity, but it is not clear what is really needed.

We can completely analyze the space of all doubling measures on F^∞ in a reasonably simple manner. To do this we need to introduce some auxiliary definitions. Let us think of F as being an "alphabet", so that we can form the set W of *words*, which is to say finite strings of elements of F. We include in W the empty word $\varnothing$. By a *child* of a word w of length k we mean a word w' of length $k+1$ whose initial k letters coincide with those of w, so that every word has exactly n children. Given an word $w \in W$ of length k, let N_w denote the *cell* in F^∞ corresponding to W, which means the set of elements of F^∞ whose first k entries coincide with w.

Suppose that we are given a family $\{\lambda_w\}_{w \in W}$ of probability measures on F, one for each word w. Then there is a unique Borel probability measure Λ on F^∞ which corresponds to this family in the natural way. To make this more explicit let us introduce a third type of object, a special class of nonnegative functions on W. Let $\mathcal{D}$ denote the collection of nonnegative functions h on W such that $h(\varnothing) = 1$ and

$$h(w) = \sum h(w_i), \tag{16.15}$$

for all $w \in W$, where the sum is taken over all the children of w.

Given a probability measure Λ on F^∞, we get an element h of $\mathcal{D}$ by setting

$$h(w) = \Lambda(N_w). \tag{16.16}$$

This function clearly satisfies the requirements above, because of the additivity of probability measures. Conversely, given such a function $h \in \mathcal{D}$, there is a unique Borel probability measure which satisfies (16.16). The existence of this measure can be obtained using the Riesz representation theorem, which is to say by using h to define a positive linear functional on continuous functions on F^∞. This linear functional is like the integral against our eventual measure Λ, but we can define it directly in the manner of a Riemann integral using h. Of course there are many ways to deal with the existence of Λ, and there is nothing special about this choice of method. To get uniqueness one can use Theorem 2.18 in [R] to conclude that any Borel probability measure on F^∞ is Borel regular, and then use the uniqueness for the measures associated to linear functionals. (This is all a bit ridiculous for such a simple space as F^∞, but it is also pleasantly brief.)

Thus we have a one-to-one correspondence between Borel probability measures on F^∞ and elements of $\mathcal{D}$. We can define also a one-to-one correspondence between families $\{\lambda_w\}_{w \in W}$ of probability measures on F and elements of $\mathcal{D}$. Given such a family $\{\lambda_w\}_{w \in W}$ we define h on W as follows. Of course each λ_w can be viewed as a nonnegative function on F for which the sum of its values is 1. We set $h(\varnothing) = 1$, and we define h on the words of length 1 – elements of F – to be the same as $\lambda_\varnothing$. We define $h(w)$ for words of longer length as follows. If $h(w)$ has been defined, and if w' is a child of w obtained by adding $e \in F$ to the end of w, then we set

$$h(w') = h(w)\,\lambda_w(e). \tag{16.17}$$

It is easy to see that this satisfies the requirements of elements of $\mathcal{D}$, and conversely that all elements of $\mathcal{D}$ arise in this manner. Different $\{\lambda_w\}_{w \in W}$'s can give the same function h, but only when one of the λ_w's vanishes somewhere. This happens if and only if $h(w) = 0$ for some word w, and it is the same as saying that the corresponding measure vanishes on some cell. If we do not allow vanishings of these types then $\{\lambda_w\}_{w \in W}$ is determined by h.

In conclusion we get a correspondence between Borel probability measures on F^∞ and families $\{\lambda_w\}_{w \in W}$ of probability measures on F, and this correspondence is one-to-one under a mild nonvanishing condition. It is not very difficult to show that $\{\lambda_w\}_{w \in W}$ gives rise to a *doubling* measure on F^∞ if and only if there is an $\epsilon > 0$ so that $\lambda_w \in \mathcal{P}_\epsilon$ for all $w \in W$. In this case we do have a one-to-one correspondence between λ's and doubling measures.

This gives a complete description of all doubling measures with total mass 1, and of course this normalization is minor.

This description admits a simpler form when the number n of elements of F is 2. In that case a probability measure on F is determined by a single number in $[0, 1]$, namely the probability of one of the elements of F, call it a. Elements of $\mathcal{P}_\epsilon$ correspond to numbers in $[\epsilon, 1 - \epsilon]$. Thus probability measures on F^∞ correspond to functions from W into $[0, 1]$, and doubling measures are in one-to-one correspondence with functions that map W into $[\epsilon, 1 - \epsilon]$ for some $\epsilon > 0$ (which is allowed to depend on the function).

In principle we might as well just work with an F having exactly two elements, because of Proposition 16.9.

Problem 16.11 *Take $F = \{0, 1\}$. Given a function $g : W \to [\epsilon, 1 - \epsilon]$ for some $\epsilon > 0$ we get a doubling measure on F^∞ as above, and then a metric $D(x, y)$ as in (16.4). We know from Proposition 16.9 that all compact Ahlfors regular uniformly disconnected metric spaces are bilipschitz equivalent to some $(F^\infty, D(x, y))$ which arises in this manner. Which g's correspond to BPI spaces? When do a pair of g's correspond to spaces which are bilipschitz equivalent? When do a pair of g's correspond to spaces which have subsets of positive measure which are bilipschitz equivalent? Or for which there is a Lipschitz mapping from a subset of one onto a set of positive measure in the other?*

There are plenty of natural questions here, once one has such a nice description of a class of spaces in this way. One can try to do "statistics" on this class, using the g's.

Note the similarity between this parameterization of uniformly disconnected Ahlfors regular spaces and the class of sets discussed in Chapter 13.

Concerning the problem stated above, notice that our simple examples of "self-similar" doubling measures on spaces F^∞ where F has more than two elements lead to more tricky examples of measures on F^∞ with $F = \{0, 1\}$ that lead to BPI spaces than the infinite products on F^∞ that we discussed before.

16.5 Reasonably self-similar pairs

We shall give more examples of doubling measures soon, but before we do that we want to give some general definitions and constructions. The point is that the next examples live naturally on the real line, which is not directly interesting for the constructions in Section 16.2, as discussed in the paragraph of (16.6). However, these examples enjoy additional structure which permits more interesting constructions. This additional structure can be formulated as in the next definition, and we shall explain the constructions afterwards.

Definition 16.12 *Let $(M, d(x,y))$ be a metric space and let μ be a doubling measure on M. We call (M, μ) a* reasonably self-similar pair *("(M,μ) is RSS" for short) if there exist constants $L_1, L_2 > 0$ so that for each pair of points $x_1, x_2 \in M$ and radii $0 < r_1, r_2 \leq \operatorname{diam} M$ there is a mapping $\phi : B(x_1, L_1 r_1) \to M$ such that $\phi(B(x_1, L_1 r_1)) \supseteq B(x_2, r_2)$, ϕ is L_2-conformally bilipschitz with scale factor r_2/r_1, and*

$$L_2^{-1} \frac{\mu(B(x_1, r_1))}{\mu(B(x_2, r_2))} \mu(A) \leq \mu(\phi(A)) \leq L_2 \frac{\mu(B(x_1, r_1))}{\mu(B(x_2, r_2))} \mu(A) \tag{16.18}$$

for all Borel sets $A \subseteq B(x_1, L_1 r_1)$.

To understand what this condition means it is helpful to forget about the measure μ for the moment. Let us call a metric space $(M, d(x,y))$ *reasonably self-similar* or RSS if it satisfies the conditions above except for (16.18). This leads to a stronger version of the BPI condition. Stronger except that we do not need to ask that M be Ahlfors regular, since we are not just asking for conformally bilipschitz mappings between general sets of substantial measure.

Note that many of our examples of BPI sets were in fact RSS. In the basic examples we did not exploit very much the fact that the BPI condition permits us to use strange subsets. This is used mildly in Sections 2.4 and 2.5, but there the subsets were still pretty special. For the general notion, however, it adds a lot of flexibility.

For the record, let us state the following question explicitly.

Problem 16.13 *Is every BPI space BPI equivalent to an RSS space?*

This is an obvious variant on the issue of what kind of "good model" one can hope to have for a BPI geometry.

To understand the notion of RSS pairs it is helpful to convert the doubling measure into a quasimetric $D(x,y)$ as in (16.4). That is, (16.18) amounts to saying that ϕ is conformally bilipschitz also with respect to $D(x,y)$, and with the natural scale factor. This is not too difficult to show. (Note that the RSS condition for $(M, d(x,y))$ implies uniform perfectness.)

In other words ϕ should be conformally bilipschitz with respect to two geometries which might be quite different.

In particular we have that $(M, D(x,y))$ is RSS as a (quasimetric) space in its own right, at least if we assume that M is uniformly perfect, so that balls for $d(x, y)$ are roughly the same as balls for $D(x, y)$, as in Lemma 16.7. The RSS property for $(M, D(x, y))$ is not so interesting if M is the real line, as noted above, but the point is that the structure of an RSS pair permits us to make other constructions.

Lemma 16.14 *Suppose that (M, μ) and (M', μ') are RSS pairs, and that M and M' are either both bounded or both unbounded. Then $(M \times M', \mu \times \mu')$ is an RSS pair.*

For this assertion we can use any of the standard metrics on the product space. It does not matter which, they are all the same to within a bounded factor.

The lemma is easy to derive from the definitions, and we omit the argument.

Suppose now that (M, μ) and (M', μ') are as in the lemma, and that M and M' are also uniformly perfect. Let $D_\times(\cdot, \cdot)$ be the quasimetric associated to $\mu \times \mu'$ as in (16.4) for some choice of α. Then, as above, we can conclude that $(M \times M', D_\times(\cdot, \cdot))$ is RSS as a quasimetric space (using also Lemma 16.7). The point now is that this space is very different from the product of $(M, D(\cdot, \cdot))$ and $(M', D'(\cdot, \cdot))$, where $D(\cdot, \cdot)$ and $D'(\cdot, \cdot)$ are associated to μ and μ' in the usual manner.

Note the subtleties of the definition of $D_\times(\cdot, \cdot)$. It is constructed from the product measure $\mu \times \mu'$, but using the notion of balls on the product $M \times M'$. This last contains a slightly tricky point that we should make emphasize. We have mentioned before that to define a measurement of distance from a doubling measure we do not really need the original background metric at full strength, but only the class of balls with respect to that metric. (In order to obtain the weakened form of the triangle inequality involved in the definition of a quasimetric we also need to have a rough idea of what it means to "double" a ball, but this is still much less than knowing the precise radii of balls, as in [P2].) When we go to products this idea encounters a difficulty, as we cannot easily determine what is a ball on a product space without knowing both what are balls in the individual spaces and what are the radii of these balls. Even if we do not care about radii for balls in the product space, we need to know about radii of balls in the factor spaces. Thus the metrics on the factors play a larger role when we make the product. It is for this reason that we need the mappings ϕ in Definition 16.12 to respect both the metric and the measure in order to make this kind of product construction. In other words, even if we are only interested in $D_\times(\cdot, \cdot)$ by the end, we need mappings on the factors to respect both the measures and the *background* metrics in order to get mappings on the product that respect $D_\times(\cdot, \cdot)$ in the right way.

We can think of the deformation of a metric through a doubling measure as being analogous to the following standard construction in Riemannian geometry. If one has a Riemannian metric on some space, and also a nonvanishing volume form (which may be independent of the metric), then there is a unique conformal deformation of the given metric whose associate volume form is the one that we started with. This new metric is analogous to the quasimetric that we get from a doubling measure and a metric. If we have a pair of Riemannian manifolds with a pair of volume forms on them, we can make the same construction on the individual spaces, or we can take the product and then make the same construction there. Taking the product first is very different from taking it after making the construction, and the difference occurs already at the level of inner products on vector spaces. One can see it in the plane, in the difference between circles and ellipses.

When making RSS spaces out of products in this manner, it is reasonable to make a fairly trivial choice for one of the factors. One could take (M', μ') to be the real line equipped with Lebesgue measure, for instance. This pair may not be exciting on its own, but in combination with an interesting (M, μ) one can get an interesting product. One might have μ being an interesting doubling measure on $\mathbf{R}$, and then the quasimetric $D_\times(\cdot,\cdot)$ on $\mathbf{R} \times \mathbf{R}$ will be interesting even though its counterparts for each factor are not so exciting.

We shall give some examples of nontrivial RSS pairs in the next sections. It is natural to wonder whether there are nontrivial examples on all RSS spaces, what kinds of general constructions exist, etc. The Heisenberg group provides an intriguing case to consider. One can also extend the definition to accommodate spaces like the ones in Section 2.5.

16.6 Riesz products

Let $\{a_j\}_{j=0}^{\infty}$ be a sequence of real numbers in $[-1, 1]$, and consider

$$\mu = \prod_{j=0}^{\infty} (1 + a_j \cos 3^j x). \tag{16.19}$$

This is called a *Riesz product.* To make sense of it we take start with the partial products

$$g_k = \prod_{j=0}^{k} (1 + a_j \cos 3^j x). \tag{16.20}$$

Each g_k is a smooth function which is 2π periodic and *nonnegative.* If one writes out $\cos x = \frac{1}{2}(\exp(ix) + \exp(-ix))$ and uses this to expand the product g_k into a trigonometric polynomial, then one sees that each term in the expansion of the product leads to a multiple of $\exp(imx)$ for an integer m that does not occur for more than one term. In other words, no integer m can be written in more that one way as a sum of numbers of the form $\pm 3^j$, where the j's are distinct integers. This is not special to 3, anything larger would work as well.

As k increases new terms are added to the Fourier expansion of g_k but none of the old terms are changed. The new terms correspond to frequencies which go to $\pm\infty$ as $k \to \infty$. Thus we get a formal Fourier expansion for the limit of the g_k's. In fact this convergence works in the sense of distributions, because the Fourier coefficients are all bounded by 1 in absolute value.

Thus the g_k's converge in the sense of distributions. The limit μ is a measure because all the g_k's are nonnegative. μ is 2π-periodic with Fourier series obtained as above, with 0th term equal to 1 in particular.

Note the similarity between this product and products of probability measures on F^∞, especially when F has two elements. This analogy can be made concrete by thinking of $F = \{-1, 1\}$ as the multiplicative group with two elements, and then a product of probability measures on F^∞ is really the same as a Riesz product, we are simply using the compact abelian group F^∞ instead of the circle, and the corresponding group of characters instead of exponentials.

As in the story of product measures on F^∞, there is a simple criterion for a Riesz product to be a doubling measure, namely

$$\sup_j |a_j| < 1. \tag{16.21}$$

This criterion is simple and well known, as in 8.8 (a) on p.40 of [St2]. See [BA] and [Se8] for related discussions. Let us mention a couple of the main points. The first is that

$$\sup_{2I} g_k(x) \leq C \inf_{2I} g_k(x) \tag{16.22}$$

whenever I is an interval of length $\leq 10 \cdot 3^{-k}$, say. This is not too difficult to prove, using (16.21) and the observation that

$$|\cos 3^j\, x - \cos 3^j\, y| \leq 2 \cdot 3^j |x - y| \tag{16.23}$$

by the mean value theorem. This quantity is small, like a geometric series, when $j < k$ and $x, y \in I$, although for j near k one is better off using the fact that each $|1 + a_j \cos 3^j\, x|$ is bounded away from 0, by (16.21). The second main point is that

$$\mu = g_k \cdot \mu_{k+1}, \tag{16.24}$$

where

$$\mu_k = \prod_{j=k}^{\infty} (1 + a_j \cos 3^j\, x). \tag{16.25}$$

This infinite product can be defined in the same manner as before. It has the nice feature that μ_k is periodic of period $2\pi/3^k$.

Thus for each k we can write μ as the product of a function which is roughly constant at the scale of 3^{-k} and a measure that is periodic of period $2\pi/3^{k+1}$. From these properties it is not too hard to derive the doubling property for μ.

One of the basic properties of Riesz products is that they are singular with respect to Lebesgue measure exactly when

$$\sum_j a_j^2 = \infty. \tag{16.26}$$

See Theorem 7.6 on p.209 of [Z].

We can get some nice self-similarity properties of Riesz products by taking the coefficient sequence to be periodic. Here is a precise statement.

Lemma 16.15 *Let $\{a_j\}_{j=0}^{\infty}$ be a periodic sequence of real numbers which satisfy (16.21), and let μ be as defined above, but viewed now as a measure on the unit circle $\mathbf{T}$. Then $(\mathbf{T}, \mu)$ is an RSS pair (Definition 16.12).*

We have already seen that μ is doubling. As for the RSS property, suppose that $\{a_j\}_{j=0}^{\infty}$ is periodic with period p. For notational convenience let us continue to work on the real line instead of the circle. Consider intervals I, I' of the form $I = [l\, 3^{-pn} 2\pi, (l+1)\, 3^{-pn} 2\pi]$, $I' = [l'\, 3^{-pn'} 2\pi, (l'+1)\, 3^{-pn'} 2\pi]$, where l, l', n, n' are integers, with n, n' nonnegative. There is an obvious affine mapping between I and I', and the restrictions of μ_{pn} and $\mu_{pn'}$ correspond under these mappings. This last assertion can be verified using the periodicity of $\{a_j\}_{j=0}^{\infty}$ and of the μ_k's. The RSS condition can be verified using this system of intervals and the affine mappings between them, and with the assistance of (16.22) to pass from information about the μ_{pn}'s to information about μ. This proves the lemma.

Riesz products are well-studied in classical analysis, and the story of quasimetrics provides the possibility of geometric interpretations of questions and results from analysis. We can take Cartesian products of Riesz products to get doubling measures on $\mathbf{R}^n$, and we can use these measures to make geometries as in (16.4). We can get RSS spaces by using periodic coefficient sequences, as discussed in Section 16.5. It is not clear what is really the geometric structure of the spaces that we obtain in this way. One can start with the obvious uniqueness questions.

Problem 16.16 *Suppose that $(\mathbf{T}, \mu)$ is an RSS pair obtained from a Riesz product with periodic coefficients as above. Let $(\mathbf{T}, dx)$ be the RSS pair obtained using Lebesgue measure. Use the product (as in Lemma 16.14) to get get a new quasimetric $D_\times(\cdot, \cdot)$ on $\mathbf{R}^2$ as in Section 16.5. When are these quasimetric spaces bilipschitz or BPI equivalent? To what extent is the coefficient sequence determined by the geometry?*

We have restricted ourselves to the sequence of frequencies $\{3^j\}$ to simplify the discussion and notation, but this is not necessary. One could even allow sequences of frequencies which are not a perfect geometric progression. It is easier to see self-similarity when one uses geometric progressions, but the general case makes sense and has interesting possibilities.

16.7 Riemann surfaces

Suppose that S_1 and S_2 are compact Riemann surfaces (without boundary) with the same genus $g > 1$. We want to produce a doubling measure using these surfaces. The book [Ah] provides a good reference for the background material that we shall need in this section.

Let $\phi : S_1 \to S_2$ be an orientation-preserving diffeomorphism between S_1 and S_2. The uniformization theorem implies that we can represent each S_i as U/Γ_i, where U denotes the upper half-plane in $\mathbf{C}$ and Γ_1, Γ_2 are Fuchsian groups acting on U. We can lift ϕ to $\Phi : U \to U$ in the usual manner, and this lifting has the property that

$$\gamma \mapsto \Phi \circ \gamma \circ \Phi^{-1} \tag{16.27}$$

defines a group isomorphism from Γ_1 onto Γ_2.

We may assume that

$$\Phi(z) \to \infty \text{ as } z \to \infty \text{ in } U, \tag{16.28}$$

because we can always conjugate Γ_1 by a Möbius transformation if necessary in order to achieve this normalization. It is well-known that $\Phi : U \to U$ is a "quasiconformal" mapping which extends to a homeomorphism on the closure of U (once we have made the preceding normalization). Let $h : \mathbf{R} \to \mathbf{R}$ denote the boundary homeomorphism.

Let μ denote the Borel measure on $\mathbf{R}$ obtained by pulling Lebesgue measure back using h, i.e.,

$$\mu(A) = |h(A)|. \tag{16.29}$$

This is a doubling measure, as in [BA]. It will be singular with respect to Lebesgue measure when $\phi : S_1 \to S_2$ is not homotopic to a conformal mapping. See [Ag]. When ϕ is homotopic to a conformal mapping h will be affine. It is well known that there is a family with $6g-6$ real parameters of pairwise inequivalent compact Riemann surfaces of genus g.

Actually, μ and h depend only on the homotopy class of ϕ. This is not hard to see, as homotopies of ϕ lead to deformations of Φ that are bounded in the hyperbolic metric, and hence unnoticeable at the boundary.

Lemma 16.17 *Under the conditions described above, $(\mathbf{R}, \mu)$ is an RSS pair (Definition 16.12).*

This is not hard to derive from the definitions, but let us be careful. Let I, I' be any pair of intervals in $\mathbf{R}$. Choose points $z, z' \in U$ so that

$$\operatorname{Im} z \approx |I|, \operatorname{dist}(z, I) \approx |I|, \operatorname{Im} z' \approx |I'|, \operatorname{dist}(z', I') \approx |I'|. \tag{16.30}$$

We wish also to require that z, z' lie on the same Γ_1-orbit, so that $\gamma(z) = z'$ for some $\gamma \in \Gamma_1$. We can do this because Γ_1 is co-compact.

If k is large enough then $\gamma(k\,I) \supseteq I'$. Fixing such a k, there is an $L > 0$ so that the restriction of γ to kI is L-conformally bilipschitz with scale factor $|I'|/|I|$. This follows from standard calculations. Keep in mind that γ is just a Möbius transformation. It is important here that k and L do not depend on the particular choices of I and I'.

The restriction of γ to kI enjoys similar properties with respect to μ. That is,

$$L^{-1}\,\frac{\mu(I')}{\mu(I)}\,\mu(A) \le \mu(\gamma(A)) \le L\,\frac{\mu(I')}{\mu(I)}\,\mu(A) \tag{16.31}$$

for all Borel sets $A \subseteq kI$, if L is chosen large enough. To see this we use (16.27) to get $\gamma_2 \in \Gamma_2$ such that

$$\gamma_2 \circ \Phi = \Phi \circ \gamma \tag{16.32}$$

on the closure of U, and in particular $\gamma_2 \circ h = h \circ \gamma$ on $\mathbf{R}$. Thus

$$\mu(\gamma(A)) = |h(\gamma(A))| = |\gamma_2(h(A))|. \tag{16.33}$$

This permits us to reduce (16.31) to showing that

$$L^{-1} \frac{|I_2'|}{|I_2|} |A_2| \leq |\gamma_2(A_2)| \leq L \frac{|I_2'|}{|I_2|} |A_2| \tag{16.34}$$

for all Borel sets $A_2 \subseteq h(kI)$, where $I_2 = h(I)$ and $I_2' = h(I')$. A_2 corresponds to $h(A)$ for A as above. In other words, we would like the restriction of γ_2 to $h(kI)$ to be L-conformally bilipschitz with scale factor $|I_2'|/|I_2|$. Notice that $h(kI)$ is contained in a bounded multiple of $h(I) = I_2$, because of the doubling property for h (as in [BA]). Set $z_2 = \Phi(z)$, $z_2' = \Phi(z')$. We have that

$$\operatorname{Im} z_2 \approx |I_2|, \ \operatorname{dist}(z_2, I_2) \approx |I_2|, \ \operatorname{Im} z_2' \approx |I_2'|, \ \operatorname{dist}(z_2', I_2') \approx |I_2'|, \tag{16.35}$$

because of (16.30) and the usual distortion estimates for Φ (as in [Ah]). We also have that

$$\gamma_2(z_2) = \gamma_2(\Phi(z)) = \Phi(\gamma(z)) = \Phi(z') = z_2'. \tag{16.36}$$

Because $h(kI)$ is contained in a bounded multiple of $h(I) = I_2$, we may conclude that the restriction of γ_2 to $h(kI)$ is L-conformally bilipschitz with scale factor $|I_2'|/|I_2|$ (if L is large enough). Remember that γ_2 is a Möbius transformation. This implies (16.34), and the lemma follows.

Once again we have an interesting class of doubling measures on the line, from which we can get doubling measures on other spaces by taking Cartesian products. We can pass from doubling measures to quasimetrics as in (16.4), and we can get new RSS spaces as discussed in Section 16.5.

It would be interesting to understand how the finer properties of the Fuchsian groups are reflected in the geometries that one can build. For instance, we can ask the following.

Problem 16.18 *Suppose that* $(\mathbf{R}, \mu)$ *is an RSS pair obtained from a mapping* $\phi : S_1 \to S_2$ *between Riemann surfaces as above, while* $(\mathbf{R}, dx)$ *is the more trivial RSS pair using Lebesgue measure. We can take the product as in Lemma 16.14, and then we get a new quasimetric* $D_\times(\cdot, \cdot)$ *on* $\mathbf{R}^2$ *as in Section 16.5. To what extent is the initial data* $\phi : S_1 \to S_2$ *determined by this quasimetric? Through bilipschitz equivalence, or BPI equivalence, for instance? How are properties of* $\phi : S_1 \to S_2$ *reflected in the looking-down relation between spaces?*

16.8 Remarks

We have talked about using doubling measures to deform geometry in general, and we have described some concrete situations where the idea can be applied, on Cantor sets and Euclidean spaces. As usual we have not said much about the examples of intermediate connectivity, such as the Sierpinski carpet, the Sierpinski gasket, or the fractal tree (as in Section 2.4). One should be able to make interesting examples on these spaces.

It would be interesting to look more carefully at the question of how easy it is to make doubling measures on a space and how this reflects the underlying geometry. The uniform disconnectedness of Cantor sets makes it particularly easy to make doubling measures on them, one can imagine that it is easier to build doubling measures on a fractal tree than on a Sierpinski carpet.

One can also consider products of snowflakes (with different snowflake parameters) or the Heisenberg group, in terms of looking for interesting doubling measures. In particular one would like to know about constructing doubling measures which deform the geometry in a nontrivial manner and for which the deformed geometry is still BPI.

On Euclidean spaces of dimension larger than 1 there are some nice rigidity properties pertaining to doubling measures. Because of Lemma 16.7 we know that a bilipschitz mapping from $(\mathbf{R}^n, |x-y|)$ to $(\mathbf{R}^n, D(x,y))$ is quasisymmetric when $D(x,y)$ comes from a doubling measure μ as in (16.4). That is, such a mapping is quasisymmetric as a map from $(\mathbf{R}^n, |x-y|)$ into itself. It is well known that quasisymmetric maps on Euclidean spaces of dimension > 1 are always absolutely continuous, so that the doubling measure μ should be absolutely continuous.

Let us be more precise. In (16.4) there is a free parameter α, but on $\mathbf{R}^n$ one should take $\alpha = 1/n$. This would make $(\mathbf{R}^n, D(x,y))$ have Hausdorff dimension n, as in Lemma 16.8, which would certainly be necessary for bilipschitz equivalence. In this case a bilipschitz mapping f from $(\mathbf{R}^n, |x-y|)$ to $(\mathbf{R}^n, D(x,y))$ would have to push Lebesgue measure forward to a measure which is bounded above and below by constant multiples of μ. This is because μ is bounded above and below by constant multiples of Hausdorff measure with respect to $D(x,y)$, by Lemmas 16.8 and 1.2. We are also thinking implicitly of (2.3), which implies that the fact that we have only a quasimetric and not a metric causes no trouble for this business of Hausdorff measure. Thus the absolute continuity of f would lead to the absolute continuity of μ.

If μ is not absolutely continuous, then we can conclude that $(\mathbf{R}^n, D(x,y))$ is not the same as $(\mathbf{R}^n, |x-y|)$ in terms of bilipschitz equivalence.

The absolute continuity of quasisymmetric mappings on Euclidean spaces also works for mappings that are defined only on a subset of Euclidean space. See [Se5].

If we define $D(x,y)$ as in (16.4) with $\alpha = 1/n$, it is natural to ask how close we come to getting a metric instead of a quasimetric. Let us call a doubling measure μ on $\mathbf{R}^n$ a *metric doubling measure* if the associated quasimetric $D(x,y)$ with $\alpha = 1/n$ is bounded from above and below by constant multiples of an actual metric. It turns out that this condition by itself implies that μ is absolutely continuous, and more. This was observed in [DS1], but the main point is that the method of Gehring [Ge] applies.

Note well that absolute continuity of a doubling measure has immediate geometric implications, e.g., that the identity as a mapping from $(\mathbf{R}^n, |x-y|)$ to $(\mathbf{R}^n, D(x,y))$ is bilipschitz on large sets. This is not hard to check, using the existence almost everywhere of Lebesgue points for locally integrable functions, for instance, if one does not mind getting bad estimates. For this statement there is nothing special about Euclidean spaces; it works anywhere, so long as α is chosen correctly. If a measure is completely singular then it means that the identity mapping is far from being bilipschitz at most points. The absolute continuity theorems for quasisymmetric mappings imply that all quasisymmetric mappings should then behave badly (in terms of bilipschitz restrictions).

At any rate, geometries on $\mathbf{R}^n$ for $n > 1$ that we get from singular doubling measures (using Riesz products or Riemann surfaces, for instance) must be quite different from Euclidean geometry. Remember that for $n = 1$ the geometries from doubling measures are equivalent to the standard Euclidean geometry, because we can build a good mapping simply by integrating the measure. We know that these geometries must be different, but it is not clear how to describe them concretely. It is not clear how to bring out the structure of the Riesz product or the mapping between Riemann surfaces in a clear way.

17

DEFORMATIONS OF BPI SPACES

It is natural to try to make a notion of "continuous families" of BPI spaces.

The idea of continuous families of geometries is very old. One can think of continuous families of Riemann surfaces, for instance. These might be realized concretely as algebraic varieties, as solution sets of some collection of polynomial equations, and one can get continuous families by varying the coefficients.

This is very algebraic and apparently very simple, but then we can pass to doubling measures as in Section 16.7 and we get something complicated and transcendental.

We can build BPI spaces from these doubling measures, taking products as in Sections 16.5 and 16.7.

Similarly we have families of BPI spaces arising from the other constructions of doubling measures. We could make BPI spaces using Riesz products with periodic coefficients, and we could deform those coefficients. We could take the analogue of Riesz products on a symbolic Cantor set F^{∞} – namely, the product of infinitely many copies of a probability measure on F – and get a continuous family of doubling measures, and hence a family of BPI spaces.

Although we have some natural examples, it is not clear what to say in general.

These examples point to a kind of rigidity for Euclidean geometry. If we deform Euclidean geometry on $\mathbf{R}^n$ using doubling measures in a nontrivial way (perhaps constructed from Riesz products or Riemann surfaces), then we have to leave the world of metrics, we have to work with quasimetrics. We saw this in Section 16.8, how "metric doubling measures" are necessarily absolutely continuous, so that the resulting geometries remain approximately Euclidean. To stay in the world of metrics we need to pass to snowflakes. Thus we are saying that Euclidean snowflakes are more easily deformed than ordinary Euclidean spaces, even among BPI spaces, because we can make nontrivial families of doubling measures.

It is not clear how to formalize these ideas in a general way that works well. Here is one choice.

Definition 17.1 *Let $\{E_t\}$, $0 \leq t \leq 1$, be a family of closed subsets of some $\mathbf{R}^n$. We say that $\{E_t\}$ is a* BPI deformation *if each E_t is Ahlfors regular and BPI, with the same dimension and uniformly bounded regularity and BPI constants, and if the family is continuous with respect to the notion of convergence given in Definition 8.1.*

We restrict to subsets of a fixed Euclidean space for simplicity. This should not be a severe restriction, modulo passing to snowflake metrics, because of Assouad's embedding theorem. In any case it is not clear that this is such a great definition anyway, and so we should not worry too much about details like this one.

The next result provides one way to construct BPI deformations.

Proposition 17.2 *Suppose that E_0 and E_1 are BPI sets in $\mathbf{R}^n$ of dimension d and that $f : E_1 \to E_0$ is regular and surjective and has BPI as a mapping (as in Definition 14.11). Define $F_t \subseteq \mathbf{R}^n \times \mathbf{R}^n$ for $0 \leq t \leq 1$ by*

$$F_t = \{(tx, f(x)) : x \in E_1\}. \tag{17.1}$$

Then $\{F_t\}$ is a BPI deformation. (Note that $F_0 = \{0\} \times E_0$ and that each F_t for $t > 0$ is bilipschitz equivalent to E_1.)

This is not hard to check using the definitions. Clearly the F_t's are closed and they depend continuously in t in the usual sense. One can check that they are also Ahlfors regular, with a constant that does not depend on t. For instance, the mapping $\phi_t : E_1 \to F_t$ defined by

$$\phi_t(x) = (t\,x, f(x)) \tag{17.2}$$

is regular with a constant that does not depend on t, as one can check using the regularity of f. The regularity of F_t can be derived as in Lemma 12.5, with a uniform bound for the constant. It remains to check that the F_t's are BPI with uniform bounds.

It will be convenient for us to use the metric

$$d((x, y), (x', y')) = \max(|x - x'|, |y - y'|) \tag{17.3}$$

on the product space $\mathbf{R}^n \times \mathbf{R}^n$. Of course this is essentially the same as the Euclidean metric, the minor differences are insignificant for something like the BPI condition.

Fix $0 < t \leq 1$, and let $z_1, z_2 \in F_t$, $0 < r_1, r_2 \leq \operatorname{diam} F_t$ be given. We can write z_i as $z_i = (t\,x_i, f(x_i))$ for some $x_i \in E_1$. Thus

$$B(z_i, r_i) \cap F_t = \{(t\,x, f(x)) : x \in E_1, |x - x_i| < t^{-1} r_i \text{ and } |f(x) - f(x_i)| < r_i\}. \tag{17.4}$$

Here we write $B(z, r)$ for balls in the product space $\mathbf{R}^n \times \mathbf{R}^n$, with respect to the metric mentioned above. Note that the x's in this set are permitted to stray far from x_i when t is small, but they are confined to the union of a bounded number of balls of radius r_i, because of the regularity of f.

Since f has BPI there is a closed set $A \subseteq E_1 \cap B(x_1, r_1)$ (in $\mathbf{R}^n$ now, rather than the product space) with $H^d(A) \geq \theta\, r_1^d$, a mapping $\phi : A \to E_1 \cap B(x_2, r_2)$, and another mapping $\psi : f(A) \to f(\phi(A))$ such that ϕ and ψ are k-conformally

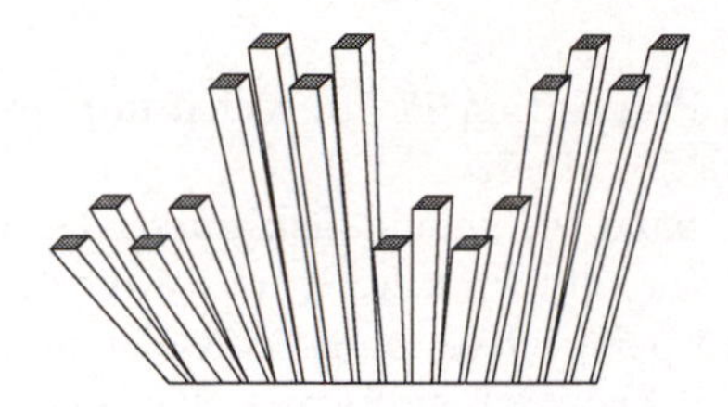

FIG. 17.1. Picture of a deformation

bilipschitz with scale factor r_2/r_1 and $\psi \circ f = f \circ \phi$ on A. Here θ and k are positive constants which depend only on $f : E_1 \to E_0$. Set

$$A_t = \{(t\,x, f(x)) : x \in A\}, \tag{17.5}$$

and define $\Phi_t : A_t \to F_t$ by

$$\Phi_t((t\,x, f(x))) = (t\,\phi(x), f(\phi(x))) = (t\,\phi(x), \psi(f(x))). \tag{17.6}$$

It is easy to see that Φ_t is k-conformally bilipschitz with scale factor r_2/r_1, because both of its coordinates enjoy this property. We have that $H^d(A_t) \geq C^{-1}\,\theta\,r_1^d$, since f is regular (Lemma 12.3). Unfortunately A_t and $\Phi_t(A_t)$ may not quite live in the correct places. If L is the maximum of 1 and the Lipschitz norm for f, then $A_t \subseteq B(z_1, L\,r_1) \cap F_t$ (in the product space now) and $\Phi_t(A_t) \subseteq B(z_2, L\,r_2) \cap F_t$. This problem can be avoided by using r_1/L, r_2/L instead of r_1, r_2 in the choice of A, ϕ, and ψ.

Thus we get that the F_t's are BPI with bounded constant when $t > 0$, and hence for all t. This completes the proof of Proposition 17.2.

As an example for the construction of the proposition, let us take E_0 to be the unit interval $[0, 1]$, and let us take E_1 to be the symbolic Cantor set F^∞ with $F = \{0, 1\}$ and with metric $d_a(\cdot, \cdot)$ as in Section 2.3 with $a = 1/2$. With this choice F^∞ is Ahlfors regular of dimension 1. It is bilipschitz equivalent to a subset of $\mathbf{R}^2$. Indeed, one can construct a model for it in the plane using the method of Section 2.5. One starts by decomposing the unit square in $\mathbf{R}^2$ into 16 squares of size 1/4. One takes as the "rule" the four corner squares. Iterating this rule as in Section 2.5 produces a subset of the plane which is bilipschitz equivalent to our F^∞.

The mapping from F^∞ onto $[0, 1]$ described in Section 11.3 is regular and BPI, and provides a concrete choice for $f : E_0 \to E_1$ in this case. (Recall that this mapping takes a binary sequence and associates to it the corresponding real number.) Using this maping we can make a deformation of BPI sets as in the proposition, and the result is roughly as depicted in Figure 17.1.

In short we can deform the Cantor set F^∞ onto $[0, 1]$. We could combine two such deformations to get a deformation from F^∞ to the Cantor set G^∞ with $G = \{1, 2, 3\}$ and with the metric from Section 2.3 which makes it regular of dimension 1. Of course we are supposed to work with these sets as subsets of the plane, say, through bilipschitz embeddings. The existence of this deformation should be compared with Proposition 11.18, which implies that neither of these Cantor sets looks down on the other.

With this example in hand we should ask ourselves whether the notion of deformation that we have is really what we want. Maybe it is too weak.

To put the matter into perspective let us consider another model situation. Consider the "universe" of compact uniformly disconnected spaces which are Ahlfors regular of dimension 1, say. We know from Proposition 16.9 that every such space is equivalent to one obtained by deforming the above metric on F^∞ ($F = \{0, 1\}$) using a doubling measure. It is reasonable to make normalizations and restrict ourselves to probability measures. This is like restricting ourselves to spaces of diameter 1. We know from Section 16.4 that we can "parameterize" the space of doubling measures on F^∞ by the set of all functions on the set of binary words which take values in $[\epsilon, 1-\epsilon]$ for some $\epsilon > 0$. This suggests natural topologies on doubling measures. We could use a "weak" topology, in which two functions are considered close if their values are close at many points (and if we can control a common ϵ), or we could use a "uniform" topology, in which two functions are close if they are uniformly close. The latter seems more natural in this context.

At any rate if we work in this world of uniformly disconnected spaces then the story of doubling measures makes it natural to confine ourselves to uniformly disconnected spaces, to not wander off to "infinity" in the space of doubling measures on F^∞. In the example above we do stray from uniformly disconnected spaces, we go towards the unit interval. This is not so unnatural either, a matter of compactifying the original noncompact space.

If we prefer the uniform topology on doubling measures, then we should be suspicious of the continuity condition in Definition 17.1. The convergence of sets used there – basically Hausdorff convergence – is more like weak convergence, it forces good approximation on many scales but not all of them at once. The uniform topology for doubling measures is like uniform approximation at all scales and locations at once. It is not clear how to impose similar conditions for sets. We want to impose approximation of geometry, and not necessarily physical location of points.

One can argue that we should not worry too much about the difference between "weak" topologies and "uniform" topologies in the context of BPI sets, that self-similarity with uniform bounds means that they are about the same.

Nonetheless it is not so clear that Definition 17.1 is a good definition, or what might be a reasonable alternative.

It would be nice to have a good notion of deformations of BPI geometries, so that one can then ask questions about the collection of BPI equivalence classes, to what extent there are many continuous families or to what extent BPI geometries are more "discrete" than that. The example above corresponds to very non-Hausdorff behavior of a certain kind of topology on the collection of BPI equivalence classes. This again makes one wonder if there is not a better notion. It makes sense that one can take a BPI space and deform it into practically anything while keeping the BPI constants bounded but not the constants of BPI equivalence bounded, as in the example. One can imagine trying to prevent this.

This problem is related to our question about finding special models for BPI geometries. One could try to prevent degenerations within a BPI equivalence class by restricting to some good class of models.

The bottom line is that there are some natural questions and examples here, but not a clear choice.

18

SNAPSHOTS

Let $(M, d(x,y))$ be a metric space which is doubling and complete, say. Given $p \in M$ and $0 < r \leq \operatorname{diam} M$ we get a "renormalized" pointed metric space $M_{p,r}$ as in Section 9.1, i.e., we get $(M, r^{-1}\, d(x,y), p)$.

It will be convenient for us to localize this more. Thus we take $S_{p,r}$ to be the closed unit ball in $M_{p,r}$. As a set this means the ball $\overline{B}(p,r)$ in M, with the metric $r^{-1}\, d(x,y)$ and basepoint p. Let us call this a *snapshot* of M, with center p and radius r.

We can think of $(p,r) \to S_{p,r}$ as a mapping from $M \times (0, \operatorname{diam} M]$ into the space of compact pointed metric spaces of radius ≤ 1 about the basepoint.

For this we should think of the upper half-space $U = M \times (0, \operatorname{diam} M]$ as being given a "quasihyperbolic" geometry. Think of something like

$$D((x,r),(y,s)) = \left|\log \frac{r}{s}\right| + \log\left(1 + \frac{d(x,y)}{\sqrt{rs}}\right). \tag{18.1}$$

The main point is that a subset Σ of U should be considered bounded when there is an $(x_0, r_0) \in U$ and a $k > 0$ such that

$$k^{-1}\, r_0 \leq r \leq k\, r_0 \quad \text{and} \quad d(x, x_0) \leq k\, r_0 \qquad \text{when } (x,r) \in \Sigma. \tag{18.2}$$

Here (x_0, r_0) is an approximate center of Σ, and k controls an approximate radius. (In terms of (18.1) the radius would be more like $\log k$ when k is large.)

This is the usual large-scale geometry to put on the upper half-space, and it fits well with the notion of a snapshot. That is, $S_{p,r}$ does not change much at small quasihyperbolic distances, but at large distances one might expect the snapshots $S_{p,r}$ to be essentially independent of each other.

This is roughly similar to the situation for harmonic functions on Euclidean upper half-spaces.

This independence of snapshots at large distances in the upper half-space comes out more clearly in some of the special "universes" that we have considered. For example, we saw in Chapter 13 how Ahlfors regular sets could be obtained from "normalized families of cubes", and that one could think of these sets as being described by a family of rules. These rules could be viewed as attached to nodes of a tree, and they could be chosen freely, modulo some normalizations if one wants to maintain Ahlfors regularity. This is very much compatible with the idea of the approximate independence of snapshots at large scales in the upper half-space; it is essentially the same idea in a simpler combinatorial context. Similarly, we can think of compact uniformly disconnected spaces which are Ahlfors regular of dimension 1 as being described completely by doubling measures on F^∞, $F = \{0, 1\}$, as in Proposition 16.9. These doubling measures can be described by functions on the set of all binary words, as in Section 16.4. One can also think of functions on the vertices of a binary tree. The values of these functions can be taken completely independently of each other, from spot to spot, and this provides another illustration of the independence of the snapshots $S_{p,r}$.

One can ask how much structure a general space should have, or how one can measure this structure. If one is thinking of something like self-similarity, one might think of measuring the amount of structure in terms of the behavior of the snapshots $S_{p,r}$, how often they are approximately the same, how close to having some kind of periodicity. If we want to think about what happens for truly general spaces, then these special universes suggest that it is like taking a collection of data, where each piece of data lives in some kind of bounded set, like a finite set of rules in the context of Chapter 13, or numbers in an interval $[\epsilon, 1-\epsilon]$ in the story of doubling measures. The pieces of data are labelled by vertices of an infinite tree, and they may be chosen independently of each other.

This is like the well-studied story of how much information is in a binary string, except that now we do not use linear "strings" but tree-like families.

Notice that there is a bit of mandatory self-similarity here, for spaces that are doubling, for instance. For doubling spaces the snapshots live in a kind of compact space. The snapshots are parameterized by the upper half-space with the quasihyperbolic metric, and we can expect something like an infinite tree of potentially independent snapshots coming from a single space. Because the snapshots lie in a compact space, though, there has to be some degree of recurrence, with some snapshots being used many times to within small errors.

In the context of self-similarity one might expect that a small collection of snapshots are approximately used a large number of times, for more "random" sets the distribution of snapshots will be more diffuse.

Considerations of statistics and entropy are obviously relevant here.

In this monograph we have studied the self-similarity condition BPI, but one could also consider general notions of self-similarity based on restrictions on the distribution of snapshots. This is a nice idea, but it is not so clear how to make it work in a nice way. A basic problem is that there does not seem to be a general and convenient compatibility between bilipschitz mappings and snapshots in general.

For uniformly rectifiable sets there is a very nice compatibility, as in [DS4]. Uniform rectifiability can be characterized in terms of almost all snapshots being approximately planes, where "almost all" is formulated in terms of the idea of *Carleson sets*. Carleson sets provide a notion of "small" subsets of upper half-spaces which is very convenient for geometry, as in [DS3, DS4, Se1] and the expositions [Se3, Se8]. It seems likely that they are useful in the BPI context as well, for saying when a collection of "bad events" is small enough to not matter too much. One could imagine applying Carleson sets in the contexts of Section 11.6 and Chapter 13, for instance.

For uniform rectifiability there is a nice rigidity phenomenon, where one can weaken the restrictions on the snapshots and get the same class of sets. For example, if most snapshots are approximately a union of planes, then the set is uniformly rectifiable, and most snapshots must in fact approximate single planes. See [DS4] for more details.

There are similar stories in more classical geometric measure theory, often expressed nowadays in terms of "tangent measures", as in [Ma].

It would be nice to have a broader understanding of the structure of sets through the distribution of snapshots.

19

SOME SETS THAT ARE FAR FROM BPI

19.1 The basic construction

Proposition 19.1 *There is a 1-dimensional regular set $F \subseteq \mathbf{R}^2$ with the property that if $A \subseteq F$ is closed, if $f : A \to F$ is Lipschitz, and $A_0 = \{x \in A : f(x) \neq x\}$, then $|f(A_0)| = 0$.*

Of course $|\cdot|$ refers to H^1 measure here.

We already know from Section 9.5 that there are regular sets whose intersection with a BPI set always has zero measure, but this is a stronger example of non-BPI behavior. Notice that the set mentioned in Proposition 19.1 has the property that it cannot have a nontrivial intersection with a CPI set (Definition 6.10) of dimension 1.

One could try to push this lack of self-similarity further into the world of weak tangents, but then one should think more about recurrence properties of snapshots, as in Chapter 18.

To prove the proposition we shall construct F as a "crank" through an iterative construction. We learned about cranks as a useful class of examples in analysis from Takafumi Murai in his work on analytic capacity. We shall use two pieces of data for this construction. The first is a listing $\{(I_j, J_j)\}_{j=1}^{\infty}$ of ordered pairs of closed and distinct dyadic intervals in the x_1-axis such that $|I_j| = |J_j|$ for each j. We ask that all ordered pairs of distinct dyadic intervals of equal length appear infinitely often in this listing. The second piece of data is a sequence $\{s_j\}_{j=1}^{\infty}$ of positive integers which satisfies

$$s_j \geq j, \tag{19.1}$$

for instance. These integers will mark how many generations are skipped at each stage of the construction, and it will be convenient to skip lots of them. Set $s_0 = 0$.

For this construction our dyadic intervals will always be closed.

We are going to take F to be the limit of a sequence of sets F_j, where the F_j's are chosen recursively. When we construct a given F_j we shall be working at the scale of $2^{-l(j)}$, where $l(j)$ is defined by $l(0) = 0$ and

$$l(j) = s_j + s_{j-1} + 5 + \max\Big(l(j-1), \log_2 \frac{1}{|I_j|}\Big) \tag{19.2}$$

when $j \geq 1$.

Take F_0 to be the x_1-axis in $\mathbf{R}^2$. In general F_j will be a countable union of closed line segments of length $\geq 2^{-l(j)}$ which are parallel to the x_1-axis. The heights of these segments will be integer multiples of $2^{-l(j)}$. Let π_1 denote the usual projection of $\mathbf{R}^2$ onto the x_1-axis. We shall construct F_j so that the segments that make up F_j are mapped onto closed dyadic intervals in the x_1-axis with disjoint interiors and which cover the x_1-axis. Thus each F_j will be nearly a graph over the x_1-axis; π_1 will be injective on F_j except on a countable set (the endpoints of the segments). Also, we shall have that

$$\operatorname{dist}(z, F_{j-1}) \leq 2^{-l(j)} \qquad \text{for all } z \in F_j. \tag{19.3}$$

If we have constructed F_j, then we shall set

$$\lambda_j(I) = \text{ the closure of } \pi_1^{-1}(I^\circ) \cap F_j \tag{19.4}$$

for each dyadic interval I in the x_1-axis, where I° denotes the interior of I. Think of this as the "lifting" of I to F_j. It is almost the same as $\pi_1^{-1}(I) \cap F_j$, but not quite; $\lambda_j(I)$ is always a subset of $\pi_1^{-1}(I) \cap F_j$, but it may miss a point in each of the fibers of π_1 over the two endpoints of I. These missing points can come from pieces of F_j to the side of I.

Let us now begin the construction of F_j in earnest. Suppose that $j \geq 1$ and that F_{j-1} has been constructed in accordance with the principles above, and let us construct F_j. Let (I_j, J_j) be as in our initial data.

Lemma 19.2 *Either $\lambda_{j-1}(I_j)$ and $\lambda_{j-1}(J_j)$ are each a single horizontal line segment, or $2^{-l(j-1)} \leq |I_j| = |J_j|$ and $\lambda_{j-1}(I_j)$, $\lambda_{j-1}(J_j)$ are unions of horizontal line segments of length $\geq 2^{-l(j-1)}$.*

This follows from our induction hypothesis.

Let us now define F_j. We shall modify $\lambda_{j-1}(J_j)$, but we shall leave the rest of F_{j-1} alone, and in particular we shall leave $\lambda_{j-1}(I_j)$ alone.

Let σ be one of the horizontal line segments in $\lambda_{j-1}(J_j)$. We replace σ with a crank $\widetilde{\sigma}$ as follows. We first decompose σ into closed line segments of length $2^{-l(j)}$ (and with disjoint interiors). We then translate each of these segments vertically by $\pm 2^{-l(j)}$, where we alternate successively between $+$ and $-$, always beginning with a $+$ at the left endpoint of σ. This gives us our crank $\widetilde{\sigma}$. We denote by $\widetilde{J}_j$ the union of the $\widetilde{\sigma}$'s, and we set

$$F_j = (F_{j-1} \backslash \lambda_{j-1}(J_j)) \cup \widetilde{J}_j. \tag{19.5}$$

Clearly F_j satisfies the structural properties discussed earlier. Repeating this process indefinitely we get a sequence of closed subsets F_j in $\mathbf{R}^2$.

Notice that $2^{-l(j)}$ is much smaller than the line segments that make up $\lambda_{j-1}(J)$ when j is large, because these segments have length equal to

$$\min(|J_j|, 2^{-l(j-1)}), \tag{19.6}$$

and the definition of $l(j)$ ensures that $2^{-l(j)}$ is much smaller than this when j is large (since $s_j \to \infty$).

One can think of F_j as being like a Lipschitz graph, except that the function being graphed is multivalued and the Lipschitz property has to be taken with respect to the dyadic geometry on the x_1-axis. More precisely, if I is a dyadic interval in the x_1-axis, then

$$\operatorname{diam} \lambda_j(I) \leq 2 \operatorname{diam} I. \tag{19.7}$$

This is not difficult to derive from the preceding construction. For small j, $\lambda_j(I)$ is likely to be just a vertical translation of I, while for large j the perturbations made above I will always be smaller than $|I|$ and will decay faster than a geometric series (because of the linear growth of the s_j's).

Lemma 19.3 *The F_j's converge (in the sense of Definition 8.1) to a closed subset F of $\mathbf{R}^2$. If I is a dyadic interval in the x_1-axis, then $\lambda_j(I)$ also converges to a compact subset of F which we denote by $\lambda(I)$. For each point p in the x_1-axis we have that $\pi_1^{-1}(p) \cap F$ consists of exactly one element when p is not an endpoint of any dyadic interval and at most two points otherwise. F is Ahlfors regular of dimension 1, and there is a 1-dimensional regular measure μ (Definition 8.27) with support equal to F whose push-forward under π_1 equals Lebesgue measure on the x_1-axis.*

The convergence of the F_j's and the $\lambda_j(I)$'s follows immediately from the construction (and the requirement that the s_j's grow at least linearly). The control on the multiplicities of π_1 is easy to derive from the "dyadic Lipschitz" condition (19.7) above. One can obtain μ as the limit of the obvious arclength measures on the F_j's, or define it more directly using Lebesgue measure on the x_1-axis, or simply take it to be 1-dimensional Hausdorff measure restricted to F. One can check that F is regular in various ways, with the help of μ if one wishes. We omit the details.

It remains to show that F satisfies the "un-self-similarity" property of Proposition 19.1. Notice that we have not used anything about the s_j's so far beyond linear growth.

Lemma 19.4 *Suppose that $A \subseteq F$ is closed and that $f : A \to F$ is Lipschitz. Let $x \in A$ be a point of density for A relative to F which satisfies $f(x) \neq x$. Then*

$$\lim_{r \to 0} \frac{|f(A \cap B(x,r))|}{r} = 0. \tag{19.8}$$

Here $|\cdot|$ refers to the natural measure μ on F as in Lemma 19.3.

Let f, A, x, be as in the lemma, and set $x_1 = \pi_1(x)$. Let $\epsilon > 0$ be given, and let I be a closed dyadic interval in the x_1-axis such that $x_1 \in 2I$. We want to show that

$$\frac{|f(A \cap \lambda(I))|}{|\lambda(I)|} < \epsilon \qquad \text{when } |I| \text{ is sufficiently small.} \tag{19.9}$$

Let J be a closed dyadic interval in the x_1-axis such that $|J| = |I|$ and $\pi_1(f(A \cap \lambda(I)))$ intersects J. We are going to show that

$$\frac{|f(A \cap \lambda(I)) \cap \lambda(J)|}{|\lambda(I)|} < \epsilon \qquad \text{when } |I| \text{ is sufficiently small.} \tag{19.10}$$

This will imply (19.9), modulo an extra constant factor in front of the ϵ (depending on the Lipschitz norm of f) which causes no trouble.

From now on we shall assume that $|I|$ is sufficiently small so that $f(x) \neq x$ implies that $I \neq J$. We shall impose further conditions on the size of I later.

Let j be one of the infinitely many positive integers such that $(I_j, J_j) = (I, J)$. In the course of the argument that follows we shall want to assume that j is large enough, depending on ϵ and the Lipschitz norm of f. Let $\widetilde{J}_j$ be as in the construction of F_j. Think of I_j, J_j as somehow "marking" part of the behavior of F_{j-1}, and then $\lambda_j(I) = \lambda_{j-1}(I)$, $\widetilde{J}_j$ reflect the new behavior in F_j. We have that

$$\begin{aligned} \mathrm{dist}(y, \lambda_j(I)) &\leq 2^{-l(j+1)+1} \qquad \text{for all } y \in \lambda(I) \\ \mathrm{dist}(z, \widetilde{J}_j) &\leq 2^{-l(j+1)+1} \qquad \text{for all } z \in \lambda(J). \end{aligned} \tag{19.11}$$

This follows easily from the construction.

The idea now is that $\lambda(J)$ has much bigger holes than $\lambda(I)$, and that this will force $f(A \cap \lambda(I)) \cap \lambda(J)$ to be small, by an argument of approximate connectedness.

Let $\{\tau_i\}$ denote the family of horizontal line segments of length $2^{-l(j)}$ which make up $\widetilde{J}_j$. Let us check that

$$\mathrm{dist}(\tau_i, \tau_m) \geq 2^{-l(j)} \qquad \text{when } i \neq m. \tag{19.12}$$

This is immediate from the construction when τ_i and τ_m come from the same parent segment in $\widetilde{J}_j$. There is a small subtlety when the parents σ and σ' of τ_i and τ_m are distinct. The only dangerous case occurs when $\pi_1(\sigma)$ and $\pi_1(\sigma')$ are adjacent dyadic intervals in the x_1-axis and τ_i and τ_m lie at endpoints of σ and σ' which are mapped to the same point by π_1. By construction all the segments which make up F_{j-1} have heights which are integer multiples of $2^{-l(j-1)}$. If the heights of σ and σ' happen to be equal, then we get the desired separation of τ_i and τ_m from the way that we defined our cranks, always starting "up" at the left-hand endpoint, and hence "down" at the right endpoint. If the heights of σ and σ' are different, then they are different by at least $2^{-l(j-1)} \geq 2^{-l(j)+5}$, and this also leads to an adequate difference in the heights of τ_i and τ_m. Thus we have (19.12).

Let $\{\alpha_n\}$ be a decomposition of $\lambda_j(I)$ into (closed) horizontal line segments (with disjoint interiors) of length $2^{-l(j)+L}$, where L is an integer, $L \in [0, s_j]$. This decomposition makes sense, because of Lemma 19.2, and because $l(j-1) \leq l(j) - L$. Set $\beta_n = \lambda(\pi_1(\alpha_n))$, so that β_n denotes the part of F which is obtained from α_n from the limit of future generations of the construction (after stage j). Observe that

$$\mathrm{dist}(y, \alpha_n) \leq 2^{-l(j+1)+1} \qquad \text{for all } y \in \beta_n, \tag{19.13}$$

by summing the usual geometric series.

Similarly, let ρ_i denote the part of F obtained from τ_i as the limit of future generations of the construction (after stage j). Again this is the same as $\lambda(\pi_1(\tau_i))$, and we have that

$$\text{dist}(y, \tau_i) \leq 2^{-l(j+1)+1} \qquad \text{for all } y \in \rho_i. \tag{19.14}$$

Combining this with (19.12) yields

$$\text{dist}(\rho_i, \rho_m) \geq 2^{-l(j)} - 2^{-l(j+1)+2} \geq 2^{-l(j)-1} \qquad \text{when } i \neq m. \tag{19.15}$$

Let G denote the set of n's such that $f(A \cap \beta_n)$ intersects at most one ρ_i, and let H denote the set of remaining n's such that $f(A \cap \beta_n)$ intersects at least two ρ_i's. Let M denote the Lipschitz constant of f.

Lemma 19.5 *If $n \in H$ and j is large enough, then $|\beta_n \backslash A| \geq C^{-1} 2^{-l(j)}$. Here C and the size requirement on j depend only on M (and not on L).*

To prove the lemma we suppose for the sake of argument that $|\beta_n \backslash A| \leq C^{-1} 2^{-l(j)}$ for a large enough C. This implies that $f(A \cap \beta_n)$ is almost connected, in the sense that

$$\Gamma = \{u \in \mathbf{R}^2 : \text{dist}(u, f(A \cap \beta_n)) \leq 2^{-l(j)-10}\} \tag{19.16}$$

is connected, for instance. This follows from the fact that α_n is a line segment, (19.13), $l(j+1) \geq l(j) + j$, and the Lipschitzness of f. We only need C and j to be large enough so that this will hold.

However, $n \in H$ means that $f(A \cap \beta_n)$ intersects at least two of the ρ_i's. We want to check that this is impossible. We know that the ρ_i's do not get too close to each other, as in (19.15). We also know that the union of all the ρ_i's is the same as $\lambda(J)$ (by definitions). Notice that $F \cap \pi_1^{-1}(J^\circ) \subseteq \lambda(J)$, where J° denotes the interior of J. The part of Γ that goes away from $\pi_1^{-1}(J^\circ)$ cannot really help to make the connection between two different ρ_i's, and inside the strip $\pi_1^{-1}(J^\circ)$ the image of f lies in $\lambda(J)$. From here it is not difficult to obtain a contradiction, and the lemma follows.

Lemma 19.6 $\mu\Big(\lambda(J) \cap \Big(\bigcup_{n \in G} f(A \cap \beta_n)\Big)\Big) \leq 2^{-L}\,|I|.$

Indeed, for each n there is an i so that $f(A \cap \beta_n) \cap \lambda(J) \subseteq \rho_i$, and ρ_i has measure $2^{-l(j)}$. Thus we have that

$$\begin{aligned} \mu\Big(\lambda(J) \cap \Big(\bigcup_{n \in G} f(A \cap \beta_n)\Big)\Big) &\leq \sum_{n \in G} 2^{-l(j)} \\ &= 2^{-L} \sum_{n \in G} 2^{-l(j)+L} \leq 2^{-L}\,|I|, \end{aligned} \tag{19.17}$$

by definition of L and the α_n's. This proves the lemma.

We are now ready to finish the proof of (19.10). Choose $L \in [0, s_j]$ to be the smallest integer such that $2^{-L} < \epsilon/2$. We can do this if s_j is large enough, and we are free to choose j as large as we want. Note that L depends only on ϵ. We obtain that

$$\mu\Big(\lambda(J) \cap \Big(\bigcup_{n \in G} f(A \cap \beta_n)\Big)\Big) < \frac{\epsilon}{2}\,|I|. \tag{19.18}$$

On the other hand,

$$\mu(f(A \cap \beta_n)) \le C\,\mu(\beta_n) \le C\,2^{-l(j)+L} \tag{19.19}$$

holds for all n, and using Lemma 19.5 we get that

$$\begin{aligned} \mu\Big(\bigcup_{n \in H} f(A \cap \beta_n)\Big) &\le C \sum_{n \in H} 2^{-l(j)+L} \\ &\le C \sum_{n \in H} 2^L\,\mu(\beta_n \backslash A) \le C\,2^L\,\mu(\lambda(I) \backslash A), \end{aligned} \tag{19.20}$$

at least if j is large enough.

We began with the assumption that x is a point of density for A in F, and that $x_1 = \pi_1(x)$ lies in $2\,I$. If $|I|$ is small enough, depending on ϵ and L (and L depends only on ϵ), then we get

$$\mu\Big(\bigcup_{n \in H} f(A \cap \beta_n)\Big) < \frac{\epsilon}{2}\,|I|. \tag{19.21}$$

Combining this with (19.18) we get (19.10), as desired.

To be a bit more careful, in this argument we start with a choice of ϵ, we then choose L and j sufficiently large, and then I sufficiently small. The Lipschitz constant of f is also involved in these choices, but that is fine, we did not go in a circle.

Thus we obtain (19.9), and Lemma 19.4 follows.

Now suppose that f, A are as in the statement of Proposition 19.1. We want to say that $|f(A_0)| = 0$. We only need to worry about the points of density of A, because the rest have measure 0. The image of the points of density of A in A_0 also has measure 0, because of Lemma 19.4 and standard covering arguments. (We did this already in Sublemma 11.8.) This completes the proof of Proposition 19.1.

19.2 Small variations on the theme

We can build other sets with amusing properties using the one above as a basic building block. Let F be as before, and let us continue to use the same notations as in the previous section. For notational convenience let us view $\mathbf{R}^2$ as being $\mathbf{C}$, identifying the x_1-axis with the real line. Set $K = \lambda([0,1])$. This is the part of F that lies above $[0,1]$, except possibly for some isolated points. It enjoys the same properties as F, except that it is compact.

Set $E_l = 2^{-l}(K + 10)$, where we are using the obvious notations for translations and dilations in $\mathbf{C}$. Put

$$E = \Big(\bigcup_{l=1}^{\infty} E_l\Big) \cup \{0\}. \tag{19.22}$$

This is a compact 1-dimensional regular set. It has some symmetry properties, which we can control very precisely.

Lemma 19.7 *For each $\epsilon > 0$ there is a closed subset A of E with the properties that $|A| < \epsilon$ and there is a finite collection of closed subsets $\{A_i\}$ of A and bilipschitz mappings $f_i : A_i \to E$ such that $E = \bigcup_i f_i(A_i)$.*

This is very simple, we take $A = (\bigcup_{l=k}^{\infty} E_l) \cup \{0\}$ for a sufficiently large k, we use dilations for the f_i's, etc. The main point is that we do not have a uniform bound on the Lipschitz constants of the f_i's.

Lemma 19.8 *For every finite M there is an $\epsilon > 0$ such that there does not exist a sequence (possibly infinite) of measurable subsets $\{A_i\}$ of E and a sequence of M-Lipschitz mappings $f_i : A_i \to E$ with the properties that*

$$\Big|\bigcup_i A_i\Big| < \epsilon \tag{19.23}$$

and

$$\Big|E \setminus \bigcup_i f_i(A_i)\Big| < \epsilon. \tag{19.24}$$

In fact one can take ϵ to be a constant multiple of M^{-1}.

Indeed, let M be given, and let $\epsilon > 0$ be small, to be chosen soon. Assume that there do exist sequences of sets and mappings $\{A_i\}$ and f_i as above, and let us seek a contradiction.

Claim 19.9 *For each $L \geq 1$,*

$$\Big|E_1 \cap \Big(\bigcup_{l=1}^{L} \bigcup_i f_i(A_i \cap E_l)\Big)\Big| \leq \sum_{l=1}^{L} 2^{l-1} \Big|\Big(\bigcup_i A_i\Big) \cap E_l\Big| \leq 2^L \epsilon. \tag{19.25}$$

The proof of the claim is simpler than the statement. The point is that if we have a Lipschitz mapping from a subset of E_l into E_1, then we can replace this mapping with the (unique) affine mapping from E_l onto E_1 that is implicit in the definition of the E_k's, modulo a set of measure 0 in the image. This follows from Proposition 19.1. This implies that

$$\Big|E_1 \cap \Big(\bigcup_i f_i(A_i \cap E_l)\Big)\Big| \leq 2^{l-1} \Big|\Big(\bigcup_i A_i\Big) \cap E_l\Big| \tag{19.26}$$

for each l, and the claim follows by summing in l and using (19.23) for the last inequality.

From the claim and (19.24) we get that

$$|E_1 \cap f_i(A_i \cap E_l)| > 0 \tag{19.27}$$

for some l such that $2^l \geq (C\,\epsilon)^{-1}$ and for some i. As before this implies that f_i must agree with our fixed affine mapping from E_l onto E_1 on a set of positive measure, because of Proposition 19.1. The Lipschitz norm of this f_i must be $\geq 2^{l-1}$. Thus a bound on the Lipschitz norms implies a lower bound on ϵ, and the lemma follows.

One can imagine making other constructions, but this illustrates the point.

20

A FEW MORE QUESTIONS

We have already mentioned a lot of questions about the looking-down relation between BPI sets, but here is another one.

Problem 20.1 *Suppose that M and N are BPI spaces of equal dimension which are not BPI equivalent (or even look-down equivalent) and that M looks down on N. Can we find a third BPI space T of the same dimension such that M looks down on T, T looks down on N, and T is not look-down equivalent to either M or N?*

Of course it is easy to generate questions once one has the look-down relation, about the structure of the collection of all look-down equivalence classes with respect to this ordering. Ideally one would like an explicit description of this ordered set.

A more basic problem is to simply have an idea of how numerous are the BPI equivalence classes, or look-down equivalence classes. These equivalence relations might be flexible enough to admit nontrivial theorems that somehow limit them. Maybe there is only a countable number of families, and these families depend on some continuous parameters in a fairly nice way, something like that.

One can ask amusing questions about Cartesian products. Here are some examples.

Problem 20.2 *Suppose that M and N are Ahlfors regular metric spaces, both bounded or both unbounded (for convenience).*

If $M \times N$ is BPI, what can we say about M and N separately?

If $M \times N$ is BPI and BPI equivalent to another such product $E \times F$, what can we conclude? Are there natural "irreducibility" conditions which permit us to conclude something like BPI equivalence of the factors?

Instead of BPI one can look at weaker conditions of self-similarity. Here is a natural concept along these lines.

Definition 20.3 (Big pieces of a fixed space) *Let M be a fixed metric space, and fix a dimension d. We let $BP(M)$ denote the class of all Ahlfors regular metric spaces of dimension d for which there exist constants $k, \theta > 0$ such that for each $x \in N$ and $0 < r \leq \operatorname{diam} N$ we can find a closed subset A of $B_N(x,r)$ with $H^d(A) \geq \theta\, r^d$ and a k-conformally bilipschitz mapping $\phi : A \to M$ with scale factor $1/r$.*

In other words, $N \in BP(M)$ if there is a substantial proportion of N inside of any given ball which looks like a subset of M modulo a conformal bilipschitz mapping. We look at all locations and scales in N, but only scale 1 in M. It would be natural to require that M be bounded also, so that all scales and locations in N are being compared to a single scale and location in M.

The strength of this condition depends on the size of M. If M is the unit ball in ℓ^∞, then all regular spaces lie in $BP(M)$, since every separable metric space admits an isometric embedding into ℓ^∞, as in (5.5). If M is compact then there are some restrictions on the class of spaces which lie in $BP(M)$. Still more restrictions ensue when M is doubling, or when it is Ahlfors regular, for instance. The case where M is Ahlfors regular of the same dimension d as in Definition 20.3 is particularly interesting, for the way in which the nonemptiness of $BP(M)$ already implies substantial self-similarity in M. That is, if we have a space N for which there is a copy of a substantial portion of M at *every* location and scale in N, then there must be many small pieces of M whose rough geometry (up to bilipschitz deformation) is replicated at unit scale in M. For instance, $BP(M)$ is empty when M is as in Proposition 19.1 (assuming that we take $d = 1$ in Definition 20.3).

One can start by looking at special cases where M is the union of two BPI spaces M_1, M_2 which are not BPI equivalent. What do elements of $BP(M)$ look like? How would the two different types of material be allowed to mix? How might this depend on the choices of M_1, M_2? It clearly matters for instance if one of M_1, M_2 looks down on the other, or if this is not the case.

Perhaps it is more interesting to consider $BP(M)$ with M somewhat larger. If M is Ahlfors regular of dimension d then elements of $BP(M)$ should be fairly restricted in their geometry. There might be interesting results concerning the relationship between the size of an M and the amount of restriction on the geometry of the elements of $BP(M)$.

This should also be compared with the story of snapshots in Chapter 18.

REFERENCES

[Ag] Agard, S. (1988). Mostow rigidity on the line: a survey. *Holomorphic functions and moduli II*, Mathematical Sciences Research Institute Publications **11**, Springer-Verlag, New York, Berlin, pp. 1–12.

[Ah] Ahlfors, L. (1966). *Lectures on Quasiconformal Mappings.* Van Nostrand mathematical studies, **10**, Van Nostrand, Princeton, NJ.

[A1] Assouad, P. (1977). *Espaces Métriques, Plongements, Facteurs.* Thèse de Doctorat, Université de Paris XI, 91405 Orsay, France.

[A2] Assouad, P. (1979). Étude d'une dimension métrique liée à la possibilité de plongement dans $\mathbf{R}^n$. *Comptes Rendus des Scéances de l'Académie des Sciences (Paris), Série I Mathématique* **288**, 731–734.

[A3] Assouad, P. (1983). Plongements Lipschitziens dans $\mathbf{R}^n$. *Bulletin de la Société Mathématique de France* **111** (1983), 429–448.

[BA] Beurling, A. and Ahlfors, L. (1956). The boundary correspondence under quasiconformal mappings. *Acta Mathematica* **96**, 125–142.

[BS] Bridson, M. and Swarup, G. (1994). On Hausdorff-Gromov convergence and a theorem of Paulin. *L'Enseignement Mathématique* **40**, 267–289.

[Ch] Christ, M. (1990). *Lectures on Singular Integral Operators. CBMS Regional Conference Series in Mathematics* **77**, American Mathematical Society, Providence, RI.

[Co] Coornaert, M. (1993). Mesures de Patterson–Sullivan sur le bord d'un espace hyperbolique au sens de Gromov. *Pacific Journal of Mathematics* **159**, 241–270.

[CP] Cooper, D. and Pignataro, T. (1988). On the shape of Cantor sets. *Journal of Differential Geometry* **28**, 203–221.

[CW1] Coifman, R. and Weiss, G. (1971). *Analyse Harmonique Non-commutative sur Certains Espaces Homogènes. Lecture Notes in Mathematics* **242**, Springer-Verlag, Berlin, New York.

[CW2] Coifman, R. and Weiss, G. (1977). Extensions of Hardy spaces and their use in analysis. *Bulletin of the American Mathematical Society* **83**, 569–645.

[D1] David, G. (1984). Opérateurs intégraux singuliers sur certaines courbes du plan complexe. *Annales Scientifique de l'École Normale Supérieure, Quatrième Série* **17**, 157–189.

[D2] David, G. (1988). Opérateurs d'intégrale singulière sur les surfaces régulières. *Annales Scientifique de l'École Normale Supérieure, Quatrième Série* **21**, 225–258.

[D3] David, G. (1988). Morceaux de graphes lipschitziens et intégrales singulières sur un surface. *Revista Matemática Iberoamericana* **4**, 73–114.

[D4] David, G. (1991). *Wavelets and Singular Integrals on Curves and Surfaces. Lecture Notes in Mathematics* **1465**, Springer-Verlag, Berlin.

[DJ] David, G. and Jerison, D. (1990). Lipschitz approximations to hypersurfaces, harmonic measure, and singular integrals. *Indiana University Mathematics Journal* **39**, 831–845.

[DS1] David, G. and Semmes, S. (1990). Strong A_∞-weights, Sobolev inequalities, and quasiconformal mappings. *Analysis and Partial Differential Equations*, C. Sadosky, editor, Lecture Notes in Pure and Applied Mathematics **122**, pp.100–111, Marcel Dekker, New York.

[DS2] David, G. and Semmes, S. (1991). *Singular Integrals and Rectifiable Sets in* $\mathbf{R}^n$*: au-delà des graphes lipschitziens. Astérisque* **193**, Société Mathématique de France.

[DS3] David, G. and Semmes, S. (1993). Quantitative rectifiability and Lipschitz mappings. *Transactions of the American Mathematical Society* **337**, 855–889.

[DS4] David, G. and Semmes, S. (1993). *Analysis of and on Uniformly Rectifiable Sets.* Mathematical Surveys and Monographs **38**, American Mathematical Society, Providence, RI.

[DS5] David, G. and Semmes, S. (1996). Uniform Rectifiability and Singular Sets. *Annales de l'Institut Henri Poincaré, Analyse Non Linéare* **13**, 383–443.

[DS6] David, G. and Semmes, S. (1996). Quasiminimal surfaces of codimension 1 and John domains. *Institut des Hautes Études Scietifiques*, preprint M/96/65, Bures-sur-Yvette, France.

[Fe] Federer, H. (1969). *Geometric Measure Theory.* Springer-Verlag, Berlin, Heidelberg, New York.

[F1] Falconer, K. (1984). *The Geometry of Fractal Sets*, Cambridge University Press.

[F2] Falconer, K. (1995). Probabilistic methods in fractal geometry. *Fractal geometry and stochastics (Finsterbergen, 1994), Progress in Probability* **37**, 3–13, Birkhäuser, Basel.

[FM] Falconer, K. and Marsh, D. (1992). On the Lipschitz equivalence of Cantor sets. *Mathematika* **39**, 223–233.

[Ge] Gehring, F. (1973). The L^p integrability of the partial derivatives of a quasiconformal mapping, *Acta Mathematica* **130**, 265–277.

[Gr1] Gromov, M. (1981). *Structures Métriques pour les Variétés Riemanniennes.* Lafontaine, J. and Pansu, P., editors. Cedic/Fernand Nathan, Paris.

[Gr2] Gromov, M. (1981). Groups of polynomial growth and expanding maps. *Institut des Hautes Études Scietifiques, Publications Mathématiques* **53**, 53–78.

[HW] Hurewicz, W. and Wallman, H. (1941). *Dimension Theory.* Princeton University Press.

[J1] Jones, P. (1989). Square functions, Cauchy integrals, analytic capacity, and harmonic measure. *Lecture Notes in Mathematics* **1384**, 24–68, Springer-Verlag, Berlin, New York.

[J2] Jones, P. (1988). Lipschitz and bi-Lipschitz functions. *Revista Matemática Iberoamericana* **4**, 115–122.

[J3] Jones, P. (1990). Rectifiable sets and the travelling salesman problem. *Inventiones Mathematicae* **102**, 1–15.

[Ki] B. Kirchheim, Rectifiable metric spaces: Local structure and regularity of the Hausdorff measure, *Proceedings of the American Mathematical Society* **121** (1994), 113–123.

[KR] Korányi, A. and Reimann, M. (1985). Quasiconformal mappings on the Heisenberg group. *Inventiones Mathematicae* **80**, 309–338.

[Lu] Luukkainen, J. (1995). Antifractal metrization. *Reports of the Department of Mathematics, University of Helsinki* **85**, Helsinki, Finland.

[LS] Luukkainen, J. and Saksman, E. Every complete doubling metric space carries a doubling measure. To appear, *Proceedings of the American Mathematical Society.*

[Ma] Mattila, P. (1995). *Geometry of Sets and Measures in Euclidean Spaces: Fractals and Rectifiability.* Cambridge Studies in Advanced Mathematics **44**, Cambridge University Press.

[Mh] Mathieu, N. (1993). Sur la non Lipschitz equivalence d'ensembles fractals générés par l.f.s. Preprint CEREMADE, Université de Paris-Dauphine.

[MS] Macias, R. and Segovia, C. (1979). Lipschitz functions on spaces of homogeneous type, *Advances in Mathematics* **33**, 257–270.

[Pe] Petersen V, P. (1993). Gromov-Hausdorff convergence of metric spaces. *Proceedings of Symposia in Pure Math Mathematics* **54**, Part 3, 489–504, American Mathematical Society, Providence, RI.

[P1] Pansu, P. (1989). Métriques de Carnot-Carathéodory et quasiisométries des espaces symétriques de rang un. *Annals of Mathematics* **129**, 1–60.

[P2] Pansu, P. (1989). Dimension conforme et sphère à l'infiniti des varieétés à courbure négative, *Annales Academiae Scientiarum Fennicae. Series A I. Mathematica* **14**, 177-212.

[R] Rudin, W. (1987). *Real and Complex Analysis.* McGraw-Hill, New York.

[Se1] Semmes, S. (1990). Analysis vs. geometry on a class of rectifiable hypersurfaces in $\mathbf{R}^n$. *Indiana University Mathematics Journal* **39**, 1005–1035.

[Se2] Semmes, S. (1993). Bilipschitz mappings and strong A_∞ weights. *Annales Academiae Scientiarum Fennicae. Series A I. Mathematica* **18**, 211–248.

[Se3] Semmes, S. (1995). Finding structure in sets with little smoothness. *Proceedings of the International Congress of Mathematicians* (Zürich, 1994), Birkhäuser, Basel, p875–885.

[Se4] Semmes, S. (1996). On the nonexistence of bilipschitz parameterizations and geometric problems about A_∞ weights. *Revista Matemática Iberoamericana* **12**, 337–410.

[Se5] Semmes, S. (1996). Quasisymmetry, Measure, and a Question of Heinonen. *Revista Matemática Iberoamericana* **12**, 727-780.

[Se6] Semmes, S. (1996). Finding Curves on General Spaces through Quantitative Topology with Applications for Sobolev and Poincaré Inequalities. *Selecta Mathematica (New Series)* **2**, 155–295.

[Se7] Semmes, S. Analysis on Metric Spaces. To appear, proceedings of the conference in honor of A.P. Calderón (Chicago, 1996).

[Se8] Semmes, S. Metric Spaces and Mappings Seen at Many Scales. To appear as an appendix to *Metric Structures in Riemannian spaces* by M. Gromov et al, Birkhäuser, Basel.

[St1] Stein, E. (1970). *Singular Integrals and Differentiability Properties of Functions.* Princeton University Press.

[St2] Stein, E. (1993). *Harmonic Analysis: Real-Variable Methods, Orthogonality, and Oscillatory Integrals.* Princeton University Press.

[Su1] Sullivan, D. (1979). The density at infinity of a discrete group of hyperbolic motions. *Institut des Hautes Études Scietifiques, Publications Mathématiques* **50**, 171–202.

[Su2] Sullivan, D. (1982). Discrete conformal groups and measurable dynamics. *Bulletin of the American Mathematical Society (New Series)* **6**, 57–73.

[TV] Tukia, P. and Väisälä, J. (1980). Quasisymmetric embeddings of metric spaces. *Annales Academiae Scientiarum Fennicae. Series A I. Mathematica* **5**, 97–114.

[Z] Zygmund, A. (1979). *Trigonometric Series*, Volumes I and II. Cambridge University Press.

INDEX